THE END OF AN EPOCH

Books by A. L. Rowse

Politics
MR. KEYNES AND THE LABOUR MOVEMENT
POLITICS AND THE YOUNGER GENERATION
THE QUESTION OF THE HOUSE OF LORDS

History
ON HISTORY
QUEEN ELIZABETH AND HER SUBJECTS
SIR RICHARD GRENVILLE OF THE *REVENGE*
TUDOR CORNWALL
THE SPIRIT OF ENGLISH HISTORY
THE USE OF HISTORY

Essays
THE ENGLISH SPIRIT

Stories
WEST-COUNTRY STORIES

Poetry
POEMS OF A DECADE, 1931–41
POEMS CHIEFLY CORNISH
POEMS OF DELIVERANCE

Autobiography
A CORNISH CHILDHOOD

THE END OF AN EPOCH

Reflections on Contemporary History

BY

A. L. ROWSE

Fellow of All Souls College, Oxford

> A man that steps aside from the world, and hath leisure to observe it without interest or design, thinks all mankind as mad as they think him for not agreeing with them in their mistakes.
>
> MARQUIS OF HALIFAX
> *Moral Thoughts and Reflections*

LONDON
MACMILLAN & CO. LTD
1947

COPYRIGHT

To
THE RIGHT HONOURABLE
ERNEST BEVIN
FELLOW WEST COUNTRYMAN
AND
A GREAT ENGLISHMAN

PRINTED IN GREAT BRITAIN

NOTE

OF the mass of political writing that I did in the years leading to the war, this book represents all that I wish to preserve. It is roughly in inverse chronological order: anyone wishing to establish the sequence of issues and events can turn to the later essays first. I hope that the book may contribute something to the understanding of contemporary history — most difficult, and least satisfying, of all forms to write. At least the book is a part of the period it deals with.

One word to explain the point of view from which it has emerged. Those years before the war were dominated by the international situation and the issues of foreign policy. I was wholly and consistently opposed — as this book shows — to the course and conduct of our foreign policy throughout those years from 1931 onwards. I felt sure in my bones that it would lead us to the disaster it did. And that aligned me, as a matter of conviction, with the Labour Party and the Left in general. This book therefore is written from the point of view of a Labour man, one who was a candidate for Parliament throughout the whole period.

But on internal policy I was much less of a believer in the economic side of Socialism. For me it meant a general control at the centre in the interests of the community, leaving the great bulk of the area to private initiative and enterprise, with social services to correct inequalities. I think I had too much common sense — plain working-class sense, if you like — to believe that people would work if you removed the incentive or the inducement. And just as I felt certain before the war that we were heading for a crash, I am as clear now that, if things go on as they are on the economic front, we are straight for a catastrophe far worse than 1931.

My complaint against politicians is that they will not sufficiently look forward into the future or make proper provision for it. It looks as if the disaster we were led into by the Conservative conduct of our foreign policy may be paralleled by the Labour Party's conduct of our economic affairs. If politicians will not look forward, perhaps that affords some justification for the reflections and admonitions of people like historians, whose business it is to take longer views.

The longer I live the more I am convinced that what matters most of all in a society is — not so much political forms and doctrines as — *responsible* government. A twofold responsibility, both upwards and downwards: of the Government towards the people and of the people towards the Government, nourishing with its currents the whole life of the State. Without that fundamental responsibility nothing can flourish or go right.

I wish to make grateful acknowledgment here to those editors who were so good as to print the protests of an independent-minded and often refractory enough contributor — of *The Times, Times Literary Supplement, Sunday Times, Observer, Economic Journal, Political Quarterly, Nineteenth Century, World Review, New Statesman, Spectator,* and of the defunct *Criterion,* edited by Mr. T. S. Eliot, who was always kind and encouraging, as was also Lord Keynes.

I cannot express what I owe to my friends, in particular Mr. G. F. Hudson and Professor Jack Simmons, for their help over this book, for their conversation and their criticism.

A. L. ROWSE

ALL SOULS COLLEGE, OXFORD
March 16, 1947

CONTENTS

		PAGE
NOTE		v
I.	APOLOGY BY WAY OF PREFACE	1
II.	BRITISH FOREIGN POLICY: A *TIMES* CORRESPONDENCE	24
III.	THE TRADITION OF BRITISH POLICY	39
IV.	REFLECTIONS ON THE EUROPEAN SITUATION	50
V.	THE END OF AN EPOCH	65
VI.	REFLECTIONS ON LORD BALDWIN	77
VII.	THE WORLD AND U.S. POLICY	90
VIII.	DEMOCRACY AND DEMOCRATIC LEADERSHIP	93
IX.	THE PROSPECTS OF THE LABOUR PARTY (1937)	103
X.	WHAT IS WRONG WITH THE CIVIL SERVICE?	118
XI.	THE DILEMMA OF CHURCH AND STATE	125
XII.	SOCIALISM AND MR. KEYNES	141
XIII.	THE RISE OF LIBERALISM	160
XIV.	THE *DÉBÂCLE* OF EUROPEAN LIBERALISM	170
XV.	WHAT IS WRONG WITH THE GERMANS	181
XVI.	GERMANY: THE PROBLEM OF EUROPE	193
XVII.	FRANCE: THE THIRD REPUBLIC	217
XVIII.	THE LITERATURE OF COMMUNISM: ITS THEORY	224
XIX.	THE THEORY AND PRACTICE OF COMMUNISM	237
XX.	MARX AND RUSSIAN COMMUNISM	253
XXI.	AN EPIC OF REVOLUTION: REFLECTIONS ON TROTSKY'S *HISTORY*	274
XXII.	QUESTIONS IN POLITICAL THEORY	291
XXIII.	MARXISM AND LITERATURE	310
INDEX		321

I
APOLOGY BY WAY OF PREFACE
[1946]

A DECADE and a half ago Lord Keynes brought together a selection of his writings on the political and economic issues from the end of the last war to the world slump of 1931. " Here are collected ", he wrote, " the croakings of twelve years, the croakings of a Cassandra who could never influence the course of events in time." We know now that that was over-modest of him: that he exercised all too much influence by his book against the Treaty of Versailles — where he was almost entirely wrong [1] — and none at all up to that time with regard to monetary policy, the return to the Gold Standard and other issues where he was quite right. How characteristic of human affairs!

I have always intended to bring together, as a kind of parallel for the later period, the political essays that I wrote during the unforgettable nightmare decade that led to the war — perhaps I should say, looking at it as an historian, the renewal of the war. I have reproduced them as a whole, rather than made selections from them, as a contribution to contemporary history. And with two purposes in mind: first, as so much evidence for my contention [2] that a point of view rooted in history can give — what people ordinarily find so difficult — a consistent line of judgment by which to interpret passing affairs; secondly, to inform the generation younger than my own, upon whom the brunt of the sacrifice fell in the war, what were the events and whose the responsibilities and shortcomings that led them into it. In effect, I publish the book as a warning. The consistency of judgment that it reveals in the midst of the welter of events themselves, and the clear estimation of the way they were going, may be read

[1] Cf. Etienne Mantoux, *The Carthaginian Peace.*   [2] Cf. my *The Use of History.*

The End of an Epoch

on page after page, year after year. It is all very well for people now to be wise after the event, after the disaster their stupidity and folly brought down upon us all, and to claim that they too understood what was happening. It was not so. It was impossible to make oneself heard in that period of confusion and humbug, of organised cant and all hypocrisy: nobody paid any attention; nobody listened. They listened to Mr. Baldwin and Mr. Chamberlain, to Sir John Simon and Sir Samuel Hoare, to *The Times* ingeminating " appeasement " and the *Daily Express* solemnly assuring its readers that they could take it there would be no war that year. That was New Year 1939.

It was a very terrible period in our history. But it must not be forgotten. It is a matter of extreme importance that we should learn its lessons, and not make its mistakes. I think this book proves that I have earned the right to be heard, above all by the younger men whom I am anxious to serve: I at any rate never misled them. The book bears its own honesty and sincerity written across it. In reading it through I find it impossible to understand now why it was all of no avail, had no effect, had not the slightest influence on anyone. But who am I to complain when the same was true of Keynes, and of that far greater Cassandra, Mr. Churchill? I had always intended to dedicate this book to Keynes, who was never anything but kind and encouraging to me. Now that he, like the great President, has fallen in the service of his country, I dedicate it to the man with whom I have always found myself in entire agreement throughout all this period, whom I have followed as a Labour candidate at Party Conferences year after year, never without being completely convinced: a fellow West Countryman whose combination of an essentially constructive mind with common-sense judgment is what I chiefly admire and find so rare in politics: the man from whom, as Foreign Secretary, the country now has most to hope.

The main purpose of this Preface must be to place the book in perspective, to put it in focus with its proper background, to correct its bias and allow for it.

Apology by Way of Preface

Here the first thing I must in justice say is that this country has completely and utterly redeemed itself by its heroic effort in the war. And secondly, that though we had our share of responsibility for the war coming about, people who are capable of taking an over-all picture of an epoch must admit that this country was not chiefly to blame. There were other countries that had a much greater responsibility for what happened to the world — and not only among the enemy countries either. The prime responsibility, of course, was Germany's and Japan's. But among the Powers that should have put a stop to their course before it was too late, our sins were chiefly sins of omission rather than of commission; our mistakes — contrary to what is usually thought of us abroad — were due to confusion of mind, a genuine shrinking among all classes from the horror of even contemplating another war, from a benevolent wish-fancy that " appeasement " might make peace possible, from a refusal to face the facts or to think clearly, from second-rateness and failure of nerve in high places. Our mistakes were not due to being over-clever, nor to a consistent confusing of means with ends, nor to a conscious Machiavellianism, as with some. The truth about Chamberlain is, not that he was a Machiavellian — as some people abroad persist in believing — but that he was an honest, conventional, self-righteous old man, a good and orderly administrator in a limited way, a definite and strong-minded character. The truth about Baldwin is that he was a kindly, sensitive business man, less honest, less clear than his successor, and that his fatal indolence well-nigh ruined his country. Disastrous as these men were as leaders of the nation, nobody can say that they were bad men: they were merely not up to the job.

But this country was at least honest, whatever our shortcomings; in the end we did not come to terms with the evil thing. We fought it, and — for a period that seemed an eternity — alone.

So much for the perspective of the outside world; so much must be said in mere fairness to the country.

But how the country redeemed itself in the tremendous

trial of the war! — it should be an inspiration to those of us who went through it, and who have any imagination, for the rest of our lives. This is not the rodomontade of a jingo: I give reasons for what I think. In the first place, it was the toughest and the closest struggle we have ever been through, a " damned near thing " — even more touch-and-go than 1588 — and a far more terrible thing that we were fighting against; in the second place, though we were fighting for our very survival, we were fighting for much more than our own survival. We were fighting for the survival of others, as before in our history; but Western civilisation depended on the outcome of this. Never had we put up such a fight for others, or in the event made such a contribution to the world. Most people just have not the imagination to see that. As an historian, I see the years 1940-45 that we have lived through as the most heroic in our history.

We certainly have redeemed our mistakes, the locust years of the decade before. And what a contrast too between the flowering of courage, energy, and genius in these years and the dreary mediocrity of the decade of disgrace. In place of a Baldwin and a Chamberlain, a Churchill; in place of an Ernest Brown, whom no one remembers now except for his local preaching and his loud voice, a Bevin; in place of a Simon and a Kingsley Wood, an Eden and an Anderson — both of whom deserved well of the State; in place of a Horace Wilson at the head of the Civil Service, the superb efficiency and self-effacement, in the interest of the war effort, of Sir Edward Bridges, son of the poet; in place of the inglorious run of the Massingberds, Ironsides, and Gorts, a Wavell, an Alexander, a Montgomery. It has been a period prolific of great men and great deeds, of much heroism on the part of simple people of which the whole country has reason to be infinitely proud. It is clear that, underneath, the stock was sound, the heart of the English people uncorrupted and unshaken.

No one could have known that so we should come through the ordeal. But this in itself is enough to account for the difference between what I wrote before the war and during it.

Apology by Way of Preface

In the writings brought together here, what concerned me with ever greater anxiety as the ordeal drew nearer year by year was that it ought not to have been necessary, that if we had followed the right course it might have been avoided, that if we had not been so disastrously misled it might not have been so nearly fatal. That was what reduced one to desperation in those years; one was almost crazy with anxiety at what was being brought down upon us, while it was quite impossible — whatever talents or gifts of clear-headedness and historical perspicuity one might have — to rouse one's countrymen, lulled as they were, befuddled and bamboozled by an immense Conservative majority, to any sense of danger until virtually too late. Mr. Churchill found as much, and he was a Conservative.

I was not a Conservative, nor am I now. In fact I always have been a Labour man, and in particular a supporter of Mr. Bevin. I must say as much to make my position clear.

The essays in this book were written from a Labour point of view, but not from one, I hope, that neglected the interest of the country. In fact, the twin foundations of Labour policy — in foreign policy, collective security (in other words the Grand Alliance), and in domestic policy a sufficient general economic control at the centre in the interest of the community as a whole (which was what I meant by Socialism) — seemed to me what the interests and the well-being of the country demanded. If it had not been, I should not have supported it. But these foundations were not only to my mind what was best for this country, but, I was convinced, what was to the general interest of Europe and the world.

So much for the point of view from which these essays were written, and to which I in general adhere.

Now for the domestic perspective, the background of home politics, the party conflicts, against which this book has to be seen.

It seems to me, looking back over it all as an historian, that both parties were divided, and the country got the worst of both worlds. The overwhelming majority of Conservatives followed Chamberlain, the business men and the party machine, who thought it possible to come to terms with Hitler and Mussolini and the Japanese: they were hopelessly wrong. But there was another wing of Conservatives, a tiny minority in Parliament — Mr. Churchill could never get more than a score of Members to follow him in the House — who were quite right about it all; and though these were a small minority, they represented the true tradition of Conservatism, the historic sense of the interests of the State as such. Similarly the Labour Party was divided. On one side there were the sensible Trade Unionists and Labour men of good sober judgment and experience, who realised that a policy of security for this country and Europe necessitated something more than words, in fact armaments. On the other hand, there was the large lunatic fringe the Labour Movement had to carry with it, of pacifists and semi-pacifists, of hopeless idealists and idealogues, and idiots like the I.L.P.

The tragedy was that the wrong-headed wing of each party played the other's game. We were treated to the disgusting spectacle of both Neville Chamberlain and George Lansbury trotting off to Hitler. The sensible elements of both parties were frustrated and kept at arm's length by the fools on either side. They came together — forced by the terrible danger of 1940 — only just in time to save the country.

But why could they not have done it before and perhaps spared us this agony? Though I was a Labour man myself, and an active Labour candidate through all these years, I was not so blinded by party spirit to the true interest of the country but that I was in favour of an approach from our side to Mr. Churchill and his supporters and to the Liberals. This book proves that here was somebody wise before the event — but a person of no importance, fighting an uphill struggle for twelve years in a hopeless part of the country from a Labour point of view, and burdened with constant illness. Though it was the right policy, indeed the only policy that offered any hope of conquering an intractable and fearfully dangerous situation, nobody paid any attention. Nobody would budge.

I suppose the truth is that I have always been too independent-minded for a party, though I was always a loyal

Apology by Way of Preface

enough Labour man. I can honestly say that all I care about now — after all that we have been through — is to look at things from the point of view of what is best for the country and for the world in general. I have lost interest in a merely sectional point of view — though I still support the Labour Party and am liable to vote Labour (perhaps out of mere conservatism!). When I think of all the sacrifices the younger generation of Englishmen have been ready to make — frustration of so many hopes, grief of spirit, anxiety, wounds, disablement, death — I cannot find it in my heart to think any other way than what I feel is for the best for us all, wherever it may lead and however hard it may be (as it often is) to bear the misunderstanding that inevitably follows from thinking independently and saying what you really think.

I know that some people regard me as a renegade and a reactionary. I know they find it difficult — particularly the middle-class intellectuals of the Left — to support the idea of a product of the working class *not* growing up in accordance with their expectations as to the way he should. Not being according to pattern, one can hardly expect to be approved of by Left orthodoxy. I am somewhat consoled by the thought that Mr. Bevin, another product of solid West-Country working people, no more meets with their approval than I do. But they can't get round either of us! In any case, one would not be worthy of one's salt as an intellectual — for I am one myself, unrepentantly — if one existed merely to repeat the clichés of Left orthodoxy, never to go outside the range of ideas made familiar by Wells and Shaw and Russell, by Lytton Strachey, the Woolfs and Bloomsbury. All these people belonged to an older generation, an easier world in which it was possible to make a great reputation by calling in question accepted ideas and institutions, pouring cheap ridicule on traditional values and standards — without understanding their inner rationale — denigrating the achievements of the past. I think it a subtler and more difficult task to understand and interpret. Anyway, *autres temps, autres mœurs de pensée*! We have other problems to deal with — more difficult and intractable too — a new line to take, different ideas to put forward.

The End of an Epoch

Enough, and more than enough, of personal apologia. Back to the decade, and the argument.

It can hardly be held against the Labour Party that they did not come to an understanding with Mr. Churchill and his supporters; for these latter were pitiably few. It would have made no difference to the immense Conservative majority determined to follow Chamberlain through thick and thin — very thick rapidly becoming very thin. Even in the vote that threw him out of power — after the disaster in Norway, the invasion of Holland, Belgium, and France now imminent — he still had a majority of blind and stupid Tories with him.

Nor can the Labour Party be held responsible for the catastrophe that nearly overwhelmed us. The meanest possible campaign was made to put the blame on the Labour Party for "not allowing us to rearm". No responsible historian could possibly allow this lie to pass. The responsibility for the disaster lies fairly and squarely upon the great Conservative Party. They held practically complete power for most of the twenty years from the end of the last war to the outbreak of this. During most of this time they had immense majorities in Parliament, though even that was perhaps less important than their virtual monopoly of economic power, their possession of practically all the means of forming opinion. Hence their long tenure of power, the constant majorities they had behind them — against the true underlying fact of our situation, namely, that the Labour Party represented the interests of the great bulk of the people. It was a very remarkable achievement in its way to have held up that natural fact from expressing itself, from translating itself into actuality. In a sense, all political observers of this period are agreed, the Conservative majorities were fictitious; they were obtained by a variety of means, not stopping short of political scares and panics — witness the disgraceful Red Letter scare election of 1924, the panic over the Gold Standard of 1931, the Baldwin Fraud of 1935 — and they were held by never allowing the true facts of the political situation to transpire. Time has had its revenge on the Tories for those twenty years. The natural thing in a democracy, unless its proper political expression is

Apology by Way of Preface

thwarted or distorted by economic power, is for the interests of property, of which the political expression is the Conservative Party, to be in a small and permanent minority. I should expect as a normal thing the Labour Party to be in a permanent majority in this country, as in countries like Australia, New Zealand, or the Scandinavian countries.

For twenty years the propertied classes in England prevented and held up that natural consummation. It was a remarkable achievement. But at what a price! — the wrapping-up of all the real political issues of those years in cotton-wool, the careful dissemination of cant and humbug which was the occupational disease of the National Government. Mr. Churchill, who was out of it all, diagnosed it, knew it for what it was worth, and realised where it was leading us: there will be evidence enough of that later in this book. His own epitaph on it appears in his rumoured comment on the Conservative submergence of 1945: " If they had had Baldwin to lead them, they wouldn't have lost the election ". How disarming that is from a man of his courage and greatness. But, of course, the answer is that if they had had Baldwin they would have lost the country.

And that was about it. Does anyone imagine that Baldwin and Chamberlain and the Conservative Party with all their resources — and they would have had the support of the Labour Party in following the right policy — could not have aroused the country to the sense of the danger it was in, if they had wanted to? The answer is transparent and simple: *They did not want to.* Baldwin, because he thought it would have a bad effect upon Conservative prospects at the election; Chamberlain, because the whole logic of his policy was to come to terms with Hitler.

Any historian must agree that the Conservative Party, so far as this country was concerned, was primarily and mainly responsible. The period covered in this book was, I should say, the worst period in the history of the party since that under George III which ended in the loss of the American Colonies. It is salutary to remember that; it must never be lost sight of or forgotten, or covered with unmerited oblivion, as a great many people would gladly forget it now. It is the duty of historians to remember for the benefit of posterity.

The electoral disaster to the Tory Party in 1945 was richly and thoroughly deserved. The British electorate, another sort of *mule du Pape*, had waited twenty years to deliver its kick. The pity of it was that it should involve Mr. Churchill, who had no responsibility for the decade of disgrace and had done his best to warn the nation throughout it. Ordinary working people who had voted Labour were sorry, with that fundamental decency that characterises them, that the defeat should seem in any way to reflect on him: many of them said as much to me on the morrow of the great defeat, and I loved them for it. There are not many countries in the world where such good feeling is possible in politics.

What of the responsibility of the Labour Party?

It was altogether secondary. During all those years it was out of power, little regarded and in a perpetual minority. No wonder it suffered so much from the disease of minority-mindedness — and still does in certain quarters. The whole purpose of the National frame-up of 1931 was to keep it permanently out of power. That was a deeply deleterious thing to the whole working of a democracy. It meant that there was virtually no Opposition in Parliament — a mere rump of 55 Labour M.P.'s, led by an old pacifist sheep in George Lansbury; all the responsible leaders of the Movement outside; no alternative Government, no possibility of one. The Tories were in clover. Is it any wonder that things went so wrong? It was the worst thing in the world for a young and immature Labour Party, that needed above all things to learn responsibility, to have such a thing done to it.

Yet the trick was performed again a third time in 1935 — 1924 was the first, 1931 the second. In 1935 it seemed the country was on the verge of a show-down with Italy over Abyssinia. The Labour Party, faced with its responsibilities in case of a war with Italy in support of the League of Nations, called a special Conference at Hastings. (I was present at it as a Labour candidate and saw what happened from close at

Apology by Way of Preface

hand.) The Conference met under the imminent danger — as it and the whole country thought — of war. It was brought face to face with its responsibilities, perfectly clearly told what they involved by Bevin and Morrison, and it clearly understood. In order to back up Baldwin's Government in the national emergency, as was its duty in spite of everything, the party threw over its own leaders, Lansbury and Cripps: Lansbury, being a pacifist, disagreed with the policy; Cripps was not a pacifist, but was convinced that Baldwin would merely take the opportunity to force an election on the country while the party was in disarray, and, having got its pledge to support the Government in backing up the League, would then go back on it.

Which was in effect what happened. Cripps proved to be right. The Labour Party turned out its parliamentary leaders in order to give the Baldwin Government all support in the event of war. Mr. Attlee was made Leader in place of George Lansbury. (I remember spending part of that evening with him alone in the hotel: no court of any kind, no cheer, a figure almost unknown to the country. Those were the circumstances in which he became Leader of the Labour Party.) Mr. Baldwin, having got his backing from the Labour Party on the score of the crisis, used his opportunity to force a snap election, pledging himself to the country to back up the League. Having got his majority, he turned his back on the League and tried to do a private deal with Mussolini through the Hoare-Laval negotiations. The country would have none of it; he was forced by public opinion to drop the attempt. But any consistent foreign policy was now in ruins, along with our good name. The effects upon Europe were utterly disastrous. Mussolini got away with it in Abyssinia, defying the British Empire and the League together. Hitler, watching, drew his own conclusions, and went forward to the militarisation of the Rhineland, the first and indispensable step in his career of external aggression. He was not opposed. The rest followed.

We ought in fact to have had that show-down with Mussolini in 1935. Italy's subsequent record in the war shows that it would not have been a thing very much to fear. The Hitler menace would not then have reached such a peak: the break in Germany would have come internally, as I hoped all along: there might have been no war. How many simple men's lives might have been saved!

We owe the whole of that to Tory policy, or lack of policy, or lack of convictions or principles. The explanation was at bottom a simple one; it was given in the remark of a Secretary of State at that time to a friend of mine as to why we did not push things to a conclusion — " We couldn't do that: it would mean that Mussolini would fall." The implication being, of course, that then Italy would fall into the hands of the Reds.

After that, coming after the experiences of 1931 and 1924, no Labour man would take anything from a Tory. I was an active member of the party all through that period and I well knew the atmosphere of complete and justified distrust. I thoroughly understood it and shared it. I would never have taken anything from Neville Chamberlain or Baldwin myself: no Labour man would. When the day of disaster came to which all this had led, in 1940, it was the Labour Party that gave Chamberlain notice to quit and insisted on Churchill as leader — the one man who had had no part in this record of trickery. By a curious irony of fate it was the Labour Party that gave Mr. Churchill his historic chance to save the country and prove himself the great man he had it in him to be. So I hope he will forgive us for the defeat we inflicted on him, against our will, in 1945.

The tragedy of all this was that after 1935 no Labour man would accept anything that came from the Tories — *even when they were right*. And this is where I criticise the Labour Party. In spite of everything, when danger threatened, we ought to have pocketed our humiliation, our pride, our distrust, everything, for the sake of the country and all that depended on it. It could not indeed be expected that the Labour Party should come out and lead the campaign for Rearmament. So far from our having any assurances as to Chamberlain's policy, his course was directly contrary to all that the interests of the country and of Europe demanded. He meant an agree-

Apology by Way of Preface

ment with Hitler, and was fool enough to believe it possible: he was not — contrary to what the Russians think — intelligent enough to realise what that involved, nor, if he had been, evil enough to have put such a thing through.

(But what of the Russians themselves? We will come to that later.)

My point of view about Rearmament differed again from my party's. In spite of our having no assurances from Chamberlain, in spite of our disgust and extreme anxiety at the way things were going, it was perfectly clear to me — as to some other Labour men, notably Dalton — that speedy Rearmament was essential and indispensable. For my part, I had realised long ago, as early as 1932, *before* Hitler came into power, that it was all up with Disarmament: 1931 was the last chance, if indeed there ever had been any. I wrote articles in Labour papers as early as that date to say as much — to be yelped at, of course, by a lot of young fools as a reactionary and perhaps a Fascist. It does not do in politics, as Mr. Churchill found, to look too far ahead; and yet, the very safety of society depends on there being some people who look further ahead than the mob.

I am prepared, then, to criticise the Labour Party legitimately for its slowness in realising what the right policy, to which it held faithfully, demanded in order to make it effective. And that was characteristic of the Labour Party all through this period: slow in the uptake, wonderfully ineffective, always making the worst of its opportunities, presenting them to its opponents on a platter, they looked like a lot of good men, struggling with adversity, poor but honest. I would have preferred them a little less honest — especially in dealing with Tories — and a great deal more effective: above all, I wanted them to have a sense of power, and not to go round with the perpetual hang-dog expression of the underdog. There is nothing more boring in the world, or in some circumstances more dangerous, than an inferiority complex. " A party that means to govern must have the mentality of the governing, not of the governed ", I kept on urging — to be greeted in the ranks of the Labour Movement either with mute incomprehension, the starry-eyed vacuity of the unteachable and the uneducable, or by the horrid suspicion that anyone who thought such a thing must be a Fascist. Still, it has been said — and 1945 would seem to warrant the saying — that the *meek* shall inherit the earth.

Writings of mine during this period, some of the essays in this book, prove that strongly critical as I was of the Conservative Party, I was not so unfair as not to criticise my own. But as an historian, I cannot let it pass that the Tories and the Labour people were *equally* to blame, were *equally* responsible for what happened. Nor can I, as an historian, lay the blame upon the people at large, as is customary now in some quarters. The truth is that the people at large are very simple and need above all things guidance, honest and sincere leadership. What is more, by their simple goodness, courage, and endurance — qualities for which we can never be grateful enough in their superb exhibition of them throughout their time of trial — they deserve the best we can give them of honest guidance and leadership, now and always, telling them the truth when things are bad, helping them to realise the dangers that impend, trusting them, not taking them in. They proved over and over, with infinite goodness throughout their long and bitter trials, that they will respond. And that is not to take an unreal or exaggerated view of their intellectual capacity. I belong to the people and I know: I do not rate their mental standards very high: the people are not intellectuals, but they have common sense and they respond to direct and honest leadership.

I am going to deal with this argument as to the responsibility of the people once and for all, in one of its highest exponents. One of the greatest of living Englishmen wrote me recently as follows:

Another matter which struck me in reading your books is your very strong criticism of Baldwin and Chamberlain. While I entirely agree with you as to their disastrous effect on our policy in the period between the wars, ought you not to recognise, as an historian, that they were fairly true representatives of the English spirit of the time, and that it was not so much the men themselves

Apology by Way of Preface

who should be criticised from a historical point of view as the lethargic spirit of the whole nation, which Churchill's eloquence and realism quite failed to rouse? It would seem to me that from the historical point of view what requires to be examined about that unhappy period is why the spirit of the nation had sunk into such ignoble ease; my view would be that Baldwin and Chamberlain were merely the representatives of that spirit and not the authors of it. Their guilt was that they knew the facts and refused to face them; but I think it is very questionable whether the nation at that time would really have given up its ease and faced the facts if their leaders had put them before them. If they would not listen to Winston, would they really have listened to Baldwin or Chamberlain?

With all deference to a truly great and heroic man, I do not agree. Churchill was at that time just an unpopular minority figure. If Baldwin and Chamberlain, in complete command of the Conservative Party, with all the resources of the National Government in power and opinion, had mobilised them behind the cry " The country is in danger ", the country would have responded. Of that I think we have proof, not only in 1940, when the country rose beyond all people's expectations of it, but long before, in 1935, when we have seen how the Labour Party responded and threw over its own leaders for the sake of national unity in time of danger. If the Tories had been willing to pursue the right policy, instead of the fatally wrong one, they would have had the Labour Party with them in their appeal to the nation. We should have had something like the unity of 1940 — and perhaps that would have prevented its fatal necessity. I do not honestly think that there is any argument about it possible. The less one has illusions about the mental level of the people, the more importance one attaches to their proper leadership. And people do not sufficiently realise that in the nature of things a democracy needs not less, but more and better, leadership.

No doubt there was something wrong with the spirit of the nation at the time, and not the spirit of this nation only. One cannot be expected to go into the diagnosis here; some of the symptoms are described later in this book, some indications given of what one inquiring observer felt while it was going on. In the end it communicated something of its fever to him: perhaps his reactions may be of use to the future historian as a kind of thermometer registering the temperature of the time. I will only venture one suggestion here: so far as this country was concerned, a great deal of the confusion was due to the underlying dilemma of the propertied classes between their class-interest and the interests of the nation. Somewhat belatedly, and very humiliatingly, they came to fit the Marxist formula; one hardly needed to be a Marxist to have the key to their behaviour, but to have been through the Marxist school was an aid to clarity.

But what, after all, of the responsibility of other countries? We must never forget that the prime responsibility lay with the Germans, and in the Far East, the Japanese. On the negative side, far more important than anything we did or did not do, was the withdrawal of the United States into isolation. It completely unsettled the balance of world politics; in Europe it left a vacuum for the Germans to walk into, the moment they recovered their spirits and their old form. The Americans had their excuse: after all, they were three thousand miles away from the European scene. We had not that excuse; but we could not carry the burden of a world peace-settlement alone.

It was not foreseen after the last war that the Americans would withdraw from their responsibilities as a world Power — and by far the strongest. As if that were not enough, the Peace settlement had already created a dangerous vacuum in Eastern Europe by the exclusion of the Russians. I have long thought that that was the real sin of the settlement of Europe after Versailles (to the Germans it was only too moderate: they needed a harsher lesson if they were ever to take it in). But excluding the Russians meant creating a thoroughly unstable situation in Eastern Europe ready for the Germans to take advantage of when they felt inclined.

I do not think that the Western Powers, including this country, were wholly to blame. After all, the Russians were in the throes of revolution; they had made their own separate

Apology by Way of Preface

peace with Germany; they were set on undermining the Western democracies by their appeal to world revolution and their support of revolutionary Communist parties. What more natural or more human than that the West should resent it? One needs a superhuman, and some people would add a supernatural, patience and understanding to do good in the realm of human affairs. We at the time, in 1918, were fighting a life-and-death struggle with Germany. When the Russian Front threatened to collapse, it was only natural that we should intervene and try to stiffen it. I think the Russians can hardly have a legitimate ground of complaint against us there.

But they certainly have a legitimate ground of complaint against our subsequent interventions, after Germany was defeated and peace made. There is no defence whatever for the Tory support of the anti-Bolshevik interventionists. It added untold suffering to the Russian people, and has to be taken very seriously into account if we are to understand the background of resentment and distrust with which the Russians regard the West. Indeed the whole record of Tory policy towards Russia since the last war has been one long story of stupidity and disaster. (Somebody ought to tell it.) It never even had the final excuse of any success. It played straight into the hands of our enemies, the Germans on one side, and the Japanese on the other, who were a much greater danger. And inside Russia it played into the hands of the worst elements, who wanted their people to be cut off from all friendly contact with the West and Western ideas of freedom and self-government. I was in agreement with Mr. Lloyd George, who from the beginning had had the hunch (perhaps Celtic intuition?) that the Bolsheviks were going to be the winners and that we should do business with them.

But that is not to say that the Russians are free of blame, are to be acquitted of all responsibility for what happened to Europe and the world.

I have said enough to show that my point of view is not at all an anti-Russian one; and the essays at the end of this book prove that I always tried to view the earlier stages of the development of the Revolution with sympathy and understanding. I devoted a great deal of study in those years to Marxism, out of which the Revolution sprang; in the 'twenties, my mind was much more occupied with the problems of Marxism and Socialist thought, by politics and economics, than it was by English history. And this study fundamentally influenced my outlook on both politics and history. Unlike most English writers on those subjects, I have been through the Marxist treadmill; and naturally I am inclined to think that to understand the world of modern politics, of contemporary history as of history in general, one needs to have been something of a Marxist.

There are my credentials: I mention them only that I may not be accused of being unfair to the Russians when I say that the upshot of the October Revolution has been a profound and bitter disappointment to the world. Where are the hopes of universal brotherhood of 1917? the promise of a fuller extension of human liberty? the expectations of greater moral freedom, of freedom of thought, the removal of barriers? In place of those hopes, what have we had? — the strengthening of nationalism, the raising of barriers, the deliberate cutting-off of contacts with the West, with what is best in Western civilisation, the belief in and practice of personal and individual freedom, freedom above all to think, for upon that one freedom all the rest depends. For want of it we have seen a most tortuous course of Machiavellian realism pursued in politics, opponents murdered in thousands, " purges " involving the lives of scores of thousands, the development of a police State where the G.P.U. does not differ in essence from the Gestapo, constant witch-hunts of heresies in the interests of a rigid and infantile orthodoxy; and what to civilised opinion is the last test, a negligible cultural output, a low-grade and materialistic civilisation that is hardly any better than Hollywood's: the way of the modern world. Where in the bad old days of the Tsars they produced Tolstoy, Turgeniev, Dostoievsky, Gogol, Tchekhov, Tchaikovsky, they now produce Gladkov's *Cement* and the symphonies of Shostakhovitch! Reflective men of my generation have been as deeply and bitterly disillusioned by the course of the Russian Revolution as ever

Apology by Way of Preface

Wordsworth, Coleridge, and Southey were by the French. The only wonder is that it has not turned us into Tories — as they were. Perhaps the record of the Tories in our time prevented that. But one thing I will say, candidly: better the society of a Tory England, with all its faults, than the brave new social order of Stalin with its gangster orthodoxy. Here one can at any rate say what one thinks and think what one likes; one does not have to subscribe to what one knows to be lies.

As for the Left intellectuals in this country who insist that a Socialist Britain should walk more closely in step with a Soviet Russia than with a capitalist America, they are making the same confusion of means with ends as the Communists with whom they travel — at a safe distance. On the fundamental issue that divides Russia from Western civilisation I must state plainly where I stand. *Communist totalitarianism does not show either in its theory or (still less) in its practice that basic respect for individual human life or for the necessary freedom for its self-expression that I regard as the keynote of Western civilisation.* And in that, Marxist orthodoxy may be found guilty of a hopeless contradiction. For economic and social institutions are but means to an end — a fuller and better life. If in pursuit of improving what are but means after all, you sacrifice the lives of millions of human beings, what is the purpose of your activity — save simply holding on to power for its own sake? A question that must sometimes occur to the mind of the successor to the Tsars in the Kremlin when he looks back over the blood-stained past and remembers his colleagues, the leaders of the October Revolution, whom he sent to their deaths. This is where the pursuit of power for its own sake leads.

There might be some excuse (though I do not hold with it) for a religious creed sacrificing men's lives in this world for some hypothetical gain in the next. But this excuse does not hold for Marxists, for whom the next world does not exist and this world is all in all. (I agree with them.) But they should attach all the more importance to the sanctity, if I may use the word, of human life, and never sacrifice lives to mere economic and social arrangements, the ends to the means. I fear they are caught out in the kind of hopeless contradiction Marxists are always attempting to point out in others.

This gives us a principle upon which to base our judgment in these matters, something to guide us in the complexity and confusion of contemporary politics. However much I might be in sympathy with the economic arrangements of Socialism — and I do not agree that the Russian economic system is a better model for the world than our own — *still, even if it were undeniably superior, its neglect of the end for the means, its failure to respect human life and freedom, condemns it.* It would be equally condemned by the Left intellectuals, if they were capable of thinking out consistently the implications of their views. They *think* they prefer Soviet Russia to capitalist America. But, in fact, they do not. They prefer life and freedom : it is the whole end and aim of their creed. They really prefer life and freedom in capitalist America to the amenities of Soviet orthodoxy as described by Koestler and George Orwell.[1] I agree with these. Whatever we may think of American capitalism, it has at least the respect for individual life and freedom which is the keynote of civilisation. From this point of view, and judged by this test, intellectual as well as practical, I do not think we can hold that Soviet Russia is, as yet, a civilised country.

I think I have proved therefore that, on their own principles, though they may be unaware of it, our Left intellectuals must prefer capitalist America to totalitarian Russia. And this apart from other imponderable considerations, which I recognise to be important — though they would not — such as, for example, that blood is thicker than water and the Americans and ourselves are fundamentally the same people.

The truth is that the Russians have not ceased to be Russians for having gone through a Revolution. And the historian appreciates best the differences that mark them off from the West. Russia has never been through the fundamental experiences that have made the West what it is : the Renaissance and Reformation with their discovery of the individual

---

[1] I recommend for reading Koestler's *Darkness at Noon, The Yogi and the Commissar,* and Orwell's *Animal Farm* : with whose point of view I am in some sympathy.

Apology by Way of Preface

and freedom of thought, the French Revolution with its achievement of civic rights. The Russians went straight from a medieval world of credulity and faith to the mechanised materialism of the twentieth century. Hence their mania for orthodoxy, the arrogant insistence upon it from the primitive and infantile to people with altogether higher mental standards.

Nor in their conduct of their external affairs, their world policy, have they justified themselves any better. In truth and fairness it must be said that their responsibility for the catastrophe is far greater than ours. It all goes back to the Revolution of 1917 and Lenin's policy of setting up a Third International, dividing the working-class movement in Europe from top to bottom. The chief victim of this madness — the madness of orthodoxy again — was the German working class: the Communists made it their main end and aim to destroy Social Democracy, and to that end even collaborated with the Nazis. To such a pass had the confusion of means with ends come! When the inevitable result of such a criminal and idiotic policy came about with the victory of the Nazis, it was Moscow that left both Communists and Socialists to their fate and made a deal with Hitler, the Soviet-German Pact, that brought on the war. In the end, in spite of "appeasement" and Mr. Chamberlain and all, we did not do such a thing, we did not come to that, we have not that responsibility. Not that it did them any good: I dare say that if the Russians had pursued a more honest and straight policy, forming a common front with the West while there was yet time, the war might never have come, the Germans been contained, and Russia herself not have had to undergo the horror of invasion. In the end, the cleverest thing in politics, as well as the best, is to be honest and straight — you do not have to be a fool.

But there seems to be a Providence that looks after fools. It is certainly amazing that we have come through as well as we have — the one great country in Europe to preserve our essential culture undamaged, its continuity unbroken. Nor have we had less luck in the Labour victory of 1945. If this country had gone Conservative at a time when the whole of Europe was going to the Left, we should have lost contact, forfeited all claims to leadership. The Communists would have captured the lead all over Europe — perhaps that is why a Labour Government in England annoys them even more than a Conservative one. *Now* there is an alternative. Perhaps we can lead the way and show the world how to achieve the advantages of a controlled economy while retaining the more precious values of individual and personal freedom. Our genius as a people always has been for a moderate course: we certainly have a better model to offer the world than either the uncontrolled capitalism of America or the morose and soulless totalitarianism of Soviet Russia.

For myself, my choice is clear. I consider the argument for a controlled economy — there never was any doubt about the right lines our foreign policy should take — as won. I spent the first part of my life advocating these objectives. Now that they are won, the emphasis of the argument changes. Now that we live under a semi-socialised economy it seems to me all the more important to keep in the public mind the values of hard work, individual initiative, the incentives of private enterprise. They are all the more important to retain in a Socialist State: for want of them it may go sagging down to inefficiency and decay. And over and above these things, it is just now in a transitional social order that we need most to understand and interpret to our people the values that made our old social order the first in Europe: public spirit, a generalised sense of public duty and willingness for public service, limitation of objectives, moderation and common sense, freedom of opportunity, *la carrière ouverte aux talents*. And, above all, I rate the creativeness of our people, our fertility in industry, science, literature, art, and thought, that are bound up with a subtly differentiated society — as if a one-class society can produce anything but a dead-level, a uniform monotony in culture. Having spent the first half of my life in a minority defending unpopular causes, I suppose I am going to spend the second half in a minority again, defending unpopular but different causes. And this in spite of the fact that I loathe being in a minority.

Apology by Way of Preface

This book, then, represents a summing-up of all that first half of my life, and of what I thought of the public issues that dominated that period. If the strain of criticism throughout is strenuous, the exacerbation acute, the anxiety sometimes verging on distraction, perhaps that is because, in the end, I have always expected more from this country than from any other.

II

BRITISH FOREIGN POLICY: A
TIMES CORRESPONDENCE

[1937]

THIS correspondence hardly needs any elucidation. My letter was written as a protest against the dangerous nonsense that was put about after Hitler's accession to power, by the supporters of Baldwin and Chamberlain, that the internal régimes of other countries were no concern of ours and had nothing to do with their foreign policy. All this, of course, in order to justify " appeasement " towards Nazi Germany. (Naturally it did not apply to Soviet Russia.) To anyone with any knowledge of history it was demonstrably untrue and deeply deleterious to the security of the country, besides going completely against the whole tradition of our policy.

But such was the confusion of mind, the humbug that was preached in season and out of season in those years, that it was almost impossible (as Mr. Churchill found) to get any clear thinking across to the public, especially in the columns of *The Times*, given up as they were to the propaganda of appeasement, of " understanding " with Hitler. The reader may note with some amusement the careful, diplomatic way in which the demonstration of its untruth had to be phrased.

I took the opportunity to indicate the true policy I thought the country ought to follow. It was the same as that being advocated across the party-barrier by Mr. Churchill. Of course it had no effect, not the slightest — though I did receive a private word of encouragement from the Foreign Office. The Foreign Office point of view was completely right during all this terrible period, but was taken no notice of, or rather turned down by Chamberlain and the Conservative Party in Parliament and throughout the country. Until too

British Foreign Policy: A Times Correspondence

late, and the disaster " appeasement " had landed us in was upon us.

The Times correspondence created a good deal of attention, and brought replies from every kind of Tory appeaser. These are mostly not worth reprinting; I have chosen only the two best representatives, the distinguished scholar Dr. Edwyn Bevan and Lord Lothian.

I note Edwyn Bevan's pathetic " Perhaps some future historian may pronounce that Mr. Rowse was right ". Of Lord Lothian I merely say that anyone who knew him knew that of his gifts of mind and personality, stability of judgment was not one. After the murders of June 30, 1934, I heard him say that the Nazis were a lot of gunmen, and the only way to deal with them was with a gun. Within a year he was in favour of " appeasement " with them and went to talk it over with Göring. But this is not to deny his great services subsequently, as Ambassador to the United States, in repairing the damage he had helped to do. As a great friend of his said to me, Lothian " died in the knowledge that he had been wrong ". He had been indeed; but not alone.

(1)

Sir,—I have followed with much interest the correspondence in your columns on the subject of mutual toleration between democracies and dictatorships, as a principle of conduct in our foreign relations. I am sure that in this country, where the free expression of all currents of opinion is happily allowed, you will welcome a statement of Labour opinion, which so far has not been heard in this discussion.

A good deal has been made of the point, both by Admiral Sir Brian Barttelot and others, that we need to be more tolerant of other forms of government than our own. The assumption is made, and has often been stated, several times by Lord Baldwin and most notably recently by the Foreign Secretary in his Leamington speech, that the internal régimes of other countries are no concern of ours in the conduct of our foreign policy.

The End of an Epoch

But surely this is very loose and superficial thinking; contrary, moreover, to the whole experience of our history? The foreign policy of a State is very largely dependent upon the character of its internal régime — may, in fact, be said to be in large part a function of it. It has happened to us again and again in the course of our experience as a nation, to find that with certain nations the very nature of their internal régime excluded mutual forbearance and toleration in the conduct of their external relations. However much other people may have desired to be on good terms with them and keep the peace of Europe, such Powers, because of the very nature of their internal régime, have refused to abide by the conditions necessary to the maintenance of mutual respect and peace.

How, then, are we to separate the two aspects of their conduct, internal and external? As a matter of fact, you have only to look at them to see that the conduct of their affairs internally bears the closest, and it may be said the most sinister, resemblance to the conduct of their affairs externally. We do not, alas, need to specify examples of such cases: the truth of the observation is written across the face of Europe and, indeed, the world today.

The point can be established from the past. It will be readily agreed that nineteenth-century liberal democracy, as an internal régime, had a corresponding foreign policy which was tolerant and on the whole pacific. In other words, upon the basis of liberal democracies some sort of peaceable international order was possible. It is doubtful, on the other hand, whether on the basis of some other sorts of internal régime such an order is possible or even desired. One does not need to make the point from the convinced and convincing disclaimers of Herr Hitler and Signor Mussolini; though the contemptuous disbelief of the one in any sort of pacific internationalism as even a political ideal is written upon every page of *Mein Kampf,* while the clarion-call of the other, " Fascism believes neither in the possibility nor the desirability of perpetual peace ", has certainly had its reverberations in the recent history of Europe and Africa. Their actions correspond with their words. And other nations, drawing their own conclusions, set about rearming too, not unnaturally. The point is, that on this view of the world, there is no possibility of any international order that can give you peace. Is not that plain from the whole recent history of Europe and the Far East?

This is the reason for our concern with the nature of the internal régime in such cases. And may I add — since it is important that the view of responsible Labour opinion should be stated in your columns above all — this is the reason for the concern, and indeed the anxiety, with which the whole opinion of the Left in this country views the foreign policy of the Government.

It is not sufficiently understood either in this country or abroad that the sheet-anchor of this country's foreign policy for centuries has been neither a hypocritical self-interest, as so many people think abroad, nor an unreal and irresponsible idealism, as some people think at home. The real safety and security of this country has lain in the fact that our legitimate interests coincided with the interests, and very often the independence, of the great number of European countries. That was our safeguard as it was theirs; it enabled us and them together to resist one threatened domination of Europe after another, first by Spain, then by France under Louis XIV and Napoleon, last of all by Germany. Until, one would have thought, it should be obvious that there is no future for Europe by the way of domination, but only a society of freely cooperating States associated together.

This view of our past policy has an obvious bearing upon the present and the future. There are so many misconceptions of our policy and role in Europe put about, as I have frequently found abroad, that it is desirable that our case should be stated. Indeed, it is doubtful if it is much understood in this country. I am myself eternally grateful to the young Canadian scholar who first opened my eyes on what I regard as the true inwardness of British policy. Superficial critics on the Continent are for ever harping on its " Machiavellian " character. This gets it completely wrong; it so happened that the interests of our own security coincided with those of other Powers that wanted to preserve their own independence. We ourselves were never sufficiently strong to challenge the very existence of other Powers; and that has been, and remains, our safeguard.

Here lies the answer to Baron von Rheinbaben and all those who disclaim any German responsibility for the late war. If one person, by his conduct of his external affairs, collects around him half a dozen antagonists, is it likely that he is innocent of all blame and that the other half-dozen are all in the wrong and had nothing to fear from his conduct?

But we on the Left have reason for anxious concern with the Government's conduct of our foreign policy in the last few years. Is it not becoming abundantly clear that the interests, and the security, of this country are bound up with those of the Left in every European country? I must not be taken to mean by the " Left " Communism. I have no wish to see Communism triumph in Europe, nor Fascism either; but there is much greater danger of the latter than the former, and it is the latter that threatens the interests of this country at every point. Conversely, if you want to find the friends of this country — and we are rapidly nearing the position when we ought to think of our friends and who they are — you will find them not on the Right but on the Left throughout the Continent.

In France, it was a Government of the Right that let us down over Abyssinia; the Popular Front Government has been much more amenable and willing to cooperate. In Poland, in Czechoslovakia, in Hungary, Rumania, Yugoslavia, in Greece, in all the countries of Central and South-Eastern Europe, it is the Right that is throwing its influence on the side of Nazi Germany or Fascist Italy. If Franco wins in Spain, will it be a good thing for this country? It is more than doubtful; almost certainly not. Then it is dangerous from the standpoint of the interests of this country to give him the measure of recognition he has received from us.

The same thing is true, and as striking, of the Great Powers. Which agrees the better with the legitimate interests and the security of this country, a Japan under the Liberal régime that prevailed up to 1931, or a Japan dominated by the Army and

British Foreign Policy: A Times Correspondence

Navy? Germany under the leadership of the Nazis, or the pacific Germany of the democratic Republic? The Italy of Mussolini and the Fascists, or the old Liberal Italy we could have called once more into being in 1935? The answer is plain which of these alternatives agrees better with international order and peace. And yet people repeat the silly cliché about foreign policy having nothing to do with other peoples' internal régimes.

What deeply disturbs the minds of all of us on the Left is that the National Government's refusal to recognise that this country's interests are bound up with those of the Left throughout Europe has led us to the point at which not only has all hope of a collective system in Europe been imperilled, but the very security of this country may be endangered.

All sensible people on the Left recognise now that we are forced to rearm, and the sooner the better for all of us. I myself have advocated rearmament, within the Labour Party, ever since 1933, when it was evident that all hope of disarmament had gone. I feel that a grave responsibility rests upon the Government of Mr. MacDonald and Mr. Baldwin for their delay to rearm in the crucial years 1933-5; Mr. Chamberlain now has to bear the burden of the consequences, which may well be dangerous.

The hesitation of the Labour Opposition to give active support to the Government's measures in regard to rearmament may be readily understood in the absence of any assurance as to their foreign policy. Indeed I have given reason for the extreme concern with which we regard that policy, and into what dangers it is leading the country. If we had more reason to have confidence in the foreign policy the Government is pursuing, there would be more ground for common agreement and active support for the measures necessary to pursue the right policy in the dangers that are before us. But perhaps those very dangers will draw both sides together to face them as before in our history.—Yours sincerely, A. L. ROWSE.

Oxford, *Aug.* 20, 1937.

The End of an Epoch

(II)

Sir,—If a majority of the Labour Party is behind Mr. A. L. Rowse in the views which his interesting and admirably straightforward letter to you today set forth, that is a fact of great significance. Mr. Rowse condemns the National Government's policy because it has not definitely taken sides with the Left in the great struggles now going on all over the world. But Mr. Rowse's letter does not tell us how he thinks that England's taking sides with the Left should have been expressed in action. Obviously it would have implied England's readiness to go to war against the dictatorships on behalf of the cause. Perhaps Mr. Rowse recognises this, for he heartily approves of our rearmament, and holds up for our example the action of our ancestors, who did go to war against the Spain of Philip II and the France of Louis XIV and Napoleon, and the action of our own generation in going to war against Germany in 1914.

Perhaps some future historian may pronounce that Mr. Rowse was right. Liberty and democracy perished from the world because at the critical moment England, the great democratic Power, the richest and strongest, hung back and would not fight. So the despots had their way all over the globe. But I think, in that case, it is unfair to blame the National Government. Their policy will have been weak and short-sighted because the temper of the British people was weak and short-sighted; no Government could rightly involve the nation in war against the nation's will. The ground of the weakness which Mr. Rowse condemns is the terrified pacifism which has possessed the nation's heart since the end of the Great War. If Mr. Rowse's view is right, what he, and his friends in the Labour Party, ought to do is to institute a wide anti-pacifist campaign over the country, persuade the nation that, in order to preserve the freedom it values, for itself and for the world, it must be prepared once more to engage in war. Then perhaps we might have a Government with a bolder foreign policy. It would not be easy, for it is especially members of Mr.

British Foreign Policy: *A* Times *Correspondence*

Rowse's party who have been active for many years past in frightening the British people of the horrors of war, in telling them that war only came because capitalists and armament makers brought it on for their selfish interests. Less than two years ago I heard an economist of the Left wing declare that in the event of any British Government trying to go to war the munition makers would paralyse its action by a strike. Since then young men of the Left have been going off to fight and die in Spain, where presumably a strike of the munition makers would not be approved.

And there is another thing which complicates the problem. A great part of the British people hating dictatorships and the suppression of free opinion as much as Mr. Rowse are not of the Left. No Government could bring the nation into a war unless it had the support of this part of the people. And to this part of the people Bolshevist Russia is an even greater horror than Nazi Germany. Great Britain could hardly plunge into the struggle against the Nazi and Fascist Powers without becoming an ally of the Bolshevist Power. Adherents of the Left may be likely to say that this horror of present-day Russia is ill-grounded; it is of course largely a question of facts and evidence about which reporters differ; but Mr. Rowse at any rate must recognise that while a large part of the British people are convinced that in Russia today there is more cruelty, more arbitrary killing, more ramifying intrigue and espionage, more suppression of opinion than in Germany, it would be difficult for a British Government to launch their people, as Russia's ally, into a world struggle for political freedom.

The truth is that the world is in such a horrible state that a clear-cut policy, such as Mr. Rowse seems to demand, is very hard to devise. It must be a choice of evils, and those who advocate any policy usually will only see the evils on the other side. Pacifists seem never to envisage squarely what it means that at a great crisis of world history England, with her immense power, should sit still and see the cause of freedom go down everywhere without striking a blow. And perhaps those who share Mr. Rowse's view do not take into account enough how very different war today would be from what it was in the days when England stood up to Philip of Spain and Louis XIV and Napoleon and the Germany of William II. —I am, Sir, your obedient servant, EDWYN BEVAN.

August 23, 1937.

(III)

Sir,—May I comment on Mr. Rowse's extremely interesting letter in *The Times* of August 23, because it sets forth a " Left " view of international affairs which many Liberal-minded people find it difficult to accept. One of the secrets of the success of British foreign policy in the past has certainly been that it has coincided with the desire of other nations to maintain their own independence. But the real secret of the old policy of the balance of power was that it was an insurance against world war, for ourselves and for others. Its essence was not that we joined one of two ideological groups and were then drawn into all its conflicts, but that we stood outside both and so were in a position either to isolate the conflict or to cast decisive weight against the side that contemplated destroying the independence of others by war. The weakness of this policy was that it only worked properly when we were so overwhelmingly powerful at sea that no State or coalition dared to challenge us. When that was not the case we were forced to join one side of the balance, and then the World War speedily followed.

Mr. Rowse wishes us to identify ourselves with a " Left " group of European Powers on the ground that all our troubles come from the militarist ambitions of the Fascist Powers. I think that is an inadequate analysis of present-day European realities. It ignores two other vital elements. The first is that Communism is at least as aggressive as Fascism, and is, in fact, the antecedent to Fascism. By the law of its being Communism seeks to overthrow liberal democracy by violence, and where this Communist violence becomes deeply rooted it inevitably produces the counter-violence of Fascism, a violence based on worship of the nation State, but Socialist in the sense that it disciplines capital no less than labour. The second is that the external aggressiveness of Fascism is in great measure the result of the maldistribution of natural resources in a world given over to economic nationalism. Russia is anti-war, not because Communism is not internationally aggressive, but because she has the greatest undeveloped area in the world — Siberia — at her disposal. The Fascist Powers are aggressive because they are both dissatisfied and powerful enough to think that by threat or use of military strength they may be able to alter the post-war *status quo* in the interest of what they regard as justice for themselves. The democracies will not save themselves by identifying themselves with either system of dictatorial violence.

But behind these immediate facts lies another — the most formidable of all — the division of Europe into an anarchy of 26 political and economic sovereignties. That was the fundamental cause of the war in 1914 — as it will be of another world war unless we can mitigate its demonic influence in time. For national sovereignty implies that international problems must be solved either by agreement or by force. There is no way of settling them, as inside a State or a federation of States, by majority decisions. Everything, therefore, that tends to inflame international disagreement makes against solution by agreement and for solution by means of war. That is why Mr. Rowse's policy — indeed the whole policy of the Left today — of emphasising ideological differences is so fatal. It does not help democracy. It only leads towards war. Ultimately the only remedy is a European federation. But Europe is not ready for the kind of solution which in 1789 ended the inter-State anarchy which was heading the United States, an area of exactly the same size as Europe, for war.

I am no admirer of the foreign policy of the National Government during the last few years, mainly because it had not the courage of its real convictions. But I think the instinct of the electorate as shown by recent by-elections, that the Government policy, at any rate at present, is less dangerous than the desire of the Left to line up the nations for another war to make the world safe for democracy, is essentially sound. For it has learned the lesson of the last war that you get neither democracy, nor liberty, nor a peace system out of a world war, however noble the ends for which it is fought. The League policy might have succeeded in the Abyssinian case if it had been pushed resolutely through, because at that stage it only involved risk of a local war. It failed because the speech of September 1935 — like the policy of the Labour Party in the Far East announced today — was a bluff without real strength and resolution to face the consequences behind it — a bluff which was successfully called by a far more experienced poker player than our National Cabinet or the Council of the League. But rearmament has destroyed the possibility of a full League policy today. In a world of armed sovereign nations your policy must be based, not on your hopes, but upon the armed strength, the resolution and the willingness for sacrifice of your own nation, and of any associates it may have in its policy.

The course both of common sense and of democratic progress today is not that proposed by Mr. Rowse, which leads inevitably towards another world war, but that recommended by the Imperial Conference. It declared against making enemies on ideological grounds, that we should rearm and that we should open discussions first with the big Powers — in whose hands gigantic rearmament has now placed the decision as to world war — and see whether frankness, good will, and a readiness to do justice, from strength and not weakness, cannot find a solution of Europe's problems, both political and economic, without war. No doubt to do this will involve a change from the happy days when the beneficiaries of Versailles could do exactly as they pleased, and certainly behaved none too wisely. The real purpose of the Rome-Berlin axis is to break the old League monopoly in favour of the *status quo* as created at Versailles. On the other hand, no General Staff today contemplates the possibility of a general war, because of the implications of aerial bombardment on industrial power and civilian morale, without the gravest anxiety. I believe therefore that, provided we are resolute and prepared to look realities in the face, a political and economic settlement with Europe based on the essential condition of the political independence of every nation within it, even though they loosely group themselves round the strong Powers, is still possible without war. It is certainly infinitely wiser to explore the possibility of agreement to the limit rather than to yield to that ideological war spirit which is, perhaps, the greatest of all the fomenters of world war at the present time.—Yours sincerely, LOTHIAN.

August 25, 1937.

(IV)

Sir,—Perhaps it may be thought only courteous to reply to some of the points to which Lord Lothian and Dr. Bevan have kindly devoted their attention.

I note their two points of agreement. Neither Lord Lothian nor Dr. Bevan has anything to say in favour of the policy which since 1931 has brought us to so dangerous a pass. Indeed what could possibly be said for it? It has led to a complete breakdown of what security there was for ourselves and others; the extraordinary series of retreats that has taken place under a " National " Government, before the threats and aggressions of the Fascist Powers, has led to a dangerous lowering of British prestige throughout the world, obvious for anybody to see, when a British Ambassador is liable to be bombed while travelling upon his lawful occasions upon the highway of the country to which he is accredited; when British ships are attacked by unrecognised rebels against a Government friendly to us; when our protests are received with the open flouts and jeers which perhaps they deserve in view of the conduct of our foreign policy in the last six years.

I note, secondly, that nobody disputes now my main contention that the internal régimes of foreign countries must concern us to some extent since their foreign policy is largely dependent upon the character of the régime. But perhaps I may make a further distinction which I hope will meet with agreement — namely, that we need to be concerned with the character of the internal régime only in so far as it affects foreign policy. And this may, I hope, resolve Lord Lothian's difficulties with regard to Communism. He tells us at one moment that Communism is as aggressive as Fascism, and then himself admits that Russia is, as he says, " anti-war ". I am glad of the admission; for the reason that it bears out the whole argument of my letter, which I was afraid Lord Lothian had not appreciated. That is that on the basis of Communism international order is possible; in fact, Communism implies internationalism. Whereas Fascism denies it; on the basis of Fascism international order and peace are impossible. Expansionism is of the very nature of Fascist régimes: you have only to look at the contemporary world to see the evidence. And the point is that it holds a terrible danger for us, since with our world-wide possessions and responsibilities we are in a most exposed position. Any foreign policy therefore that encourages or aids the Fascist Powers goes clean contrary to the interests of this country.

It is not only anti-patriotic, but it is also anti-international; since the interests of this country coincide with those of the great majority of European countries. That was the sheet-anchor of our security in the past, and should be the guide to our foreign policy in the future; it is largely because of our departure from our traditional policy since 1931 that we have fallen into the mess in which we are now floundering. It is all very well for Lord Lothian to put forward the suggestion of agreement with the Fascist Powers as an alternative to Left policy. It is not an alternative at all, for it is not even possible, except on their terms. Nor would you get agreement, even if you gave way all along the line to them, for they do not believe in international order. Their own will, their own interests, are their only law.

There is only one way to deal with this, to organise an overwhelming preponderance of Powers whose interest in peace is challenged, so that the aggressors shall not break through. For there is going to be no peace for Europe or the world if they do. Such a preponderance existed in 1931; it has been betrayed and broken by the most alarming series of retreats in our history.

It seems to me that the policy I advocate should not be

condemned as a merely party or " Left " policy. I should have thought it had much more claim to be regarded as truly national, and in the tradition that has hitherto carried us through to such success as a nation. It seems to me to be in line, for example, with Mr. Churchill's conception of what our policy should be.

One misconception runs through both Lord Lothian's and Dr. Bevan's letters. They would, I am sure, recognise that it would be sheer misrepresentation to say that the Left wanted us to line up for another war. The whole object of its policy is to give the maximum chance of maintaining peace ; and we believe that the National Government has immensely increased the chances of war by weakening the collective system, by not seeing the Disarmament Conference through to success while there was still time, by a whole series of unexpected blunders and retreats which has landed us in a position where nobody takes any notice of what we say or want.

The urgent question is, What are we to do now ? I have been asked by several writers what I propose we should do in detail. Well, I have not the resources of the Foreign Office behind me, nor am I Foreign Secretary. I would to goodness I were! But I would suggest that the least we might do, to put it negatively, is not to weaken the hands of our friends (as with " non-intervention " in Spain) and strengthen those of Powers whose aggressive designs now in full development all over the world threaten both our interests at every point and the interests of world peace. There are ways and means of advancing the interests of our friends without involving ourselves in war. I do not think I need be more explicit. A little Machiavellianism in pursuance of our own interests and those of our friends would not be at all a bad thing. Only the Government has not even got to the point of recognising who our friends are.

Positively, then, I advocate a policy which is both more patriotic and more international, since our interests are bound up as in the past with those of the great number of European countries, and with a system of collective security. If the Government would only follow, as it has not done hitherto, a

The End of an Epoch

more truly national policy, I believe it could get a large measure of support from the whole country for the rearmament necessary to strengthen its hand in maintaining this country's security and peace in Europe.—Yours sincerely, A. L. ROWSE.

September 1937.

III

THE TRADITION OF BRITISH POLICY

[1939]

At this time, when circumstances and Herr Hitler are bringing us up against the fundamental facts of the relations between this country and others, the foundations of British policy, it is very necessary that the British people should think clearly about them. And yet it is doubtful if they at all realise the full strength of their historic case. Indeed it is certain that they do not. In the years since the last war, and particularly in these last years before the present one, there has been a far too generous readiness on the part of many Englishmen to swallow German criticisms of our traditional policy through the centuries, quite unthinkingly. A real knowledge of our history, and not merely a superficial interpretation of it such as is current in some quarters on the Continent, would have shown how little these criticisms were justified; and clear thinking would usually have shown that their strongest root was in jealousy of our success as a nation, in *schadenfreude* — a word for which there is, by the way, no English equivalent.

Partly in consequence of their intensive propaganda, upon which the Germans at least have spent a good deal of money, and partly as the result of our mental laziness in thinking out our position, there has gained some currency even among ourselves, a view of our traditional policy which is at once by no means favourable and not at all justified by history. In fact, we have no reason to be ashamed of, or apologetic for, the long centuries of success which adherence to the fundamental line of our policy has brought us; when we have failed in the past it has been precisely when we have departed from that line or over-stepped it. It is ridiculous to suppose, seriously, that that long record of success has been due to " English gold ", or our " Machiavellianism ", or our diabolical cunning in working the balance-of-power system against our enemies.

The End of an Epoch

One could well wish, as an Englishman, that we had a little more cunning, a spot more Machiavellianism; and we could all do with a little more gold. But, in truth, the success of British policy in the past, which has so concerned Continental and not merely German commentators (them it has mostly exasperated out of all sense), must obviously be due, it is only reason to suppose, to something much more fundamental than such considerations.

What these reasons are that really account for our success in the past, I hope to suggest; but I should say that I owe them to an intelligent Canadian of my acquaintance, who first put them to me. It may be that there are some advantages in viewing the growth of British policy and the position of Britain in the European scene objectively from that distance and in perspective. For the rest, a true view must rest upon a careful study of our history in the modern period, from the sixteenth century onwards.

But first let us take the hostile view as it is expressed in its latest and most powerful form by Herr Hitler. It is not that we expect from him a considered or even a competent judgment as to the course we have pursued historically: it would need an historian to offer that. An autodidact himself, an autodidact of genius, even, for politics, he would not be qualified to give a judgment of any value on the history of British policy — though *Mein Kampf* has some shrewd *aperçus* on certain aspects of our politics. That consideration naturally does not deter one who aspires to lay down the law in matters of literature, art, morals, philosophy, biology, no less than in politics. Herr Hitler's characterisation of British policy in his self-justificatory message to the German nation on the outbreak of war may be taken as typical, crude and over-simplified though it is, of that hostile view which, based on history as it is supposed to be, a knowledge of history controverts.

Let us examine it. He says:

England has for centuries pursued the aim of rendering the peoples of Europe defenceless against the British policy of world-conquest by proclaiming a balance of power, in which England claimed the right to attack on threadbare pretexts and destroy that

The Tradition of British Policy

European state which at the moment seemed most dangerous. Thus at one time she fought the world power of Spain, later the Dutch, then the French, and, since the year 1871, the German.[1]

We do not, as we have said, need to take Herr Hitler as historian very seriously; it is only too obvious that his will be propaganda-history. But his view of the role of Great Britain in modern Europe needs only one question to show its one-sidedness and falsity: has it never struck him, and those who share his view, that if Britain has been successful in resisting and defeating those Powers which were so strong as to challenge the security and independence of others in Europe, it was not so much due to the superior virtues of Great Britain as to *the fact that her interest coincided with the interest of the bulk of those other, smaller Powers who were threatened*? Of course British policy has had to provide for the interest and security of the nation; there is nothing surprising in that: every nation has to provide for its own interest and security. It would be very disingenuous of Nazi leaders who believe in that and nothing else as a rule of State, who indoctrinate their people with the teaching that the interest of their own State is the alpha and omega of politics, that there are no principles higher than that, and that beyond that there is no rule of law in international affairs, no question of justice and right, but only the arbitrament of force, the rule of violence and fraud and treachery, and that everything is justified in the pursuit of power: it is disingenuous in the extreme for such people to complain that British policy in the past has been conceived in the interest of Britain and to ensure her safety.

The difference between this country and theirs is that it so happens that our interest has been the same as that of the bulk of other European countries. That is why they have been ready to join with us in coalitions and alliances throughout our modern history: because their security and independence were threatened by an overwhelmingly strong Power, often more directly and dangerously than we were ourselves, as is the case with Poland today. Not all the gold in the world, let alone the fantastic resources of Machiavellian cunning with

[1] Published in the English newspapers of September 4, 1939.

The End of an Epoch

which we are credited, would have gained us the allies we have had in our history, if it had not been to their interest to fight with us. The simple truth is, providentially for us, that the interest of this nation and that of the great bulk of Europe are one and the same, as against that of any Power so strong as to challenge the freedom and independence of the others. It is this that has been the sheet-anchor of our security in turn, the guarantee of our long record of success as a nation. When we have adhered to this line, it has guided us through great dangers to safety; it is when we have departed from it, or over-stepped it, that we have failed.

The point may be proved historically. In the course of the sixteenth century, at the beginning of our career as a modern nation, it was Spain that rose to a position of such overwhelming power and predominance as to threaten the independence, and in some cases the very existence, of other States. Portugal, for example, was overwhelmed in 1580 and the Portuguese Empire annexed to the vast Spanish Empire which already spanned the world. The Portuguese did not relish the extinction of their independence; somehow or other — it may be very unreasonable of them — peoples do not, witness the Poles and the Czechs; and the Portuguese kept alive their sentiment of independence until they won it back in 1640. Nor were they alone in being threatened: there were the Dutch, whose fight for existence is one of the epics of European history. France, too, was seriously endangered by Spanish power, not to mention the Italian States, which were too weak to resist, though they hated it. In reality all these States were more directly endangered by Spain than Elizabeth's England was. It was no less to their interest, but rather more, to join with us in checking Spanish aggression. The recurrent dilemma of British policy is perspicaciously expressed by the long-sighted, clear-headed Burghley (would that we had had a Burghley in these last few years!) in a memorandum before the outbreak of the war with Spain: should we wait until Spain had dealt singly with her opponents, had quenched Dutch independence and divided France, and was then free to turn her undivided attention to us, the last, not the

The Tradition of British Policy

first, citadel of freedom in Europe; or should we join issue while we had allies to fight with us in a common cause, our own freedom and the freedom of Europe?

The situation has been essentially the same with France in the seventeenth and eighteenth centuries, as with Germany in the twentieth. No one would dispute, I fancy, not even a French historian — certainly not a German — that the enormous preponderance of France from Louis XIV to Napoleon was a standing threat to the liberty and independence of others: in particular to Belgium, Holland, the Rhineland, Savoy, and at times to a wider circle of Powers less strong than France, to Spain, Austria, and, let it be noticed, even to Prussia. Great Britain had a common interest with them in resisting aggression, in preventing them from being overrun — as Holland was wantonly and without warning by Louis XIV in 1672, as Belgium and Holland were by the French Revolutionary armies in 1793-4, as almost the whole of Germany was by Napoleon, as Prussia was in 1806 and Spain in 1808. No wonder other countries could be got to form coalitions with us, time after time, to resist these onslaughts: it was even more directly their interest than ours. I do not go so far as to say that our own action was prompted by altruism; I content myself with affirming the obvious bond of common interest which bound us and the bulk of other European States against the aggressor.

Nor is the case at all dissimilar with Germany in this century. Herr Hitler is of course quite wrong in dating the turning-point in Anglo-German relations to 1871: he would be, since he knows no history, or has only a politically motivated smattering of it. As a matter of fact, we ought to have been on our guard about Germany from 1871, and again from the moment Hitler came into power in 1933; but we were not. All historians know that the turning-point in our relations with Germany only came with the opening of the twentieth century, with the hostility displayed towards this country during the Boer War, the rejection of feelers for an Anglo-German understanding, the determination of the German Government to challenge our sea-power by building a great fleet. Herr Hitler

The End of an Epoch

himself has roundly condemned the course of pre-1914 German policy in antagonising Great Britain. Now he tells the German people in his proclamation: " We ourselves have been witnesses of the policy of encirclement which has been carried on by England against Germany since before the war " (*i.e.* of 1914). The German propaganda-lie about encirclement has been judicially and coolly exposed by Professor Brierly in his Oxford pamphlet on " Encirclement ". It is sufficient to say here that if you antagonise everybody, threaten some people, blackmail others, conduct your policy by mingled violence and fraud, need you be surprised if the others come together to defend themselves against you? It is hardly likely that that Power which is engaged alone in a struggle with several other Powers is itself entirely in the right and they all in the wrong; it is more likely on reflection that there is something in its conduct and character which is responsible for the situation in which it finds itself.

So at least this country found from its own experience at the one period in our history when we were so strong as to be a threat to others. During the short time between the close of the Seven Years' War in 1763 and the outbreak of the American War of Independence, we were, with our lien upon the whole North American continent, far too strong (and no doubt correspondingly arrogant) in relation to other European Powers. And it produced a European coalition against us, very rightly, which defeated us and forced us to give up the American Colonies. Most Englishmen since, and some of the best Englishmen at the time, thought it a very good thing that we should have been defeated in 1782. I wish as many Germans could see that it was equally right and proper that they should have lost their pre-1914 ascendancy (very few of them have done). Their post-1871 ascendancy was not a good thing for Europe. Even at its best and comparatively responsible under Bismarck, it was designed to restrict the liberty of action and independence of France as a nation. Once he threatened to let loose a preventive war on her; he kept her perpetually in terror. And what Germany was like under his more neurotic and pathological successors (if ever there were a gang of men

The Tradition of British Policy

more unfitted to govern Europe than William II, Holstein, Bülow, Tirpitz, it is only Hitler, Göring, Goebbels, Himmler: what an *embarras de choix* the Germans do present us with, to be sure); what pre-1914 Germany was for Europe we all know, with its designs for Mittel-Europa, its oppression of the Poles, its encouragement of Austria-Hungary to subjugate the Slavs of South-Eastern Europe. We all hear so much about the oppressiveness of the Versailles Treaty — which actually in its territorial clauses was all too fair — that we are apt to forget the positive evidence we have of what sort of treaty the German Imperialists would have imposed if they had won the war. Right up to within sight of the end, they would not give up their hold on Belgium; they intended to annex the coal-producing area of north-eastern France and help themselves to a share of the French colonial empire; as for the treaties imposed on Russia and Rumania, they showed that the Germans meant to extend their empire over most of Eastern Europe. If Herr Hitler finds that there is consistency in British policy today with that before 1914, it is because there is a still more striking consistency between his Germany and the Germany of William II. What difference there is was in favour of the latter; for as all Europe knows, Nazi Germany is more unreliable, more brutal and barbaric with its system of internal repression and torture, its unprovoked onslaughts upon the independence of Austria, Czechoslovakia, and now Poland. Whatever disadvantages Versailles imposed upon Germany — and after all, they asked for it: militarists and nationalists who believe in that sort of international order have no right to complain and whine when the game goes against them: only internationalists, Liberal and Socialist, have the right to criticise peace settlements on the Versailles model, for they have something better to offer — however much, then, the Germans repine at Versailles, we have to remember that it gave liberty to more than 70 million Slavs. That was its historic achievement, which we are liable to overlook in sympathising with the (now mythical) sorrows of Germans over it.

Herr Hitler pipes to us the *leit-motiv* of *Mein Kampf*: " I have many times offered England and the English people the understanding and friendship of the German people. . . . I have always been repelled ", he weeps. It is a case too much resembling the Walrus and the Carpenter:

> " I weep for you," the Walrus said :
> " I deeply sympathise."
> With sobs and tears he sorted out
> Those of the largest size,
> Holding his pocket-handkerchief
> Before his streaming eyes.

The way to the understanding and friendship of the English people was for Herr Hitler to have been a better European: that way is now irrevocably closed to him — was closed from the very beginning by the nature of his régime, as some of us saw. It was not for want of trying to make him a better European that the British Government was driven to the conclusion it has arrived at, namely, that nobody can trust the word of the German Führer any more. Indeed, the criticism of one-half of the British electorate has been all along that their Government was prepared to go to almost any length in concessions to Herr Hitler, if only he would keep the peace and abide by the rule of law in Europe. That very excess of patience and conciliatoriness, which we of the Left in this country criticised all along, strengthens the moral case of the British Government before Europe now: it is so very obvious that they did all they could to avoid a breach, and that Hitler's outpourings about British war-inciters plotting to encircle Germany from the moment the Nazis came in is so much nonsense, unworthy of the intelligence of even the German people. That is what we *ought* to have done, alas, but did not do. No doubt there could have been an understanding with the Nazis on the basis of a share-out —

> " O Oysters," said the Carpenter,
> " You've had a pleasant run !
> Shall we be trotting home again ? "
> But answer came there none—
> And this was scarcely odd, because
> They'd eaten every one.

The Tradition of British Policy

But this procedure is not the British conception of international policy appropriate to the twentieth century.

A further criticism of British policy which we have heard frequently of late years from the Nazis by way of justifying their rapacious (they call it dynamic) conduct, is that ours was no better in the past and that the British Empire was built up in the same way. As an historian I should say that that is almost wholly untrue, besides being utterly unhistorical in its way of judging. As if the twentieth century in its code of conduct, its standards and manners, is the same as the sixteenth, or even the eighteenth century for that matter! One would have thought that human society had effected a good deal of progress since then — as indeed it has, in almost every sphere, except that these barbarians want to throw us back into a previous age, with its altogether more primitive standards and more brutal conditions. However, as a matter of fact, it is not true to say that the British Empire on the whole was built up in this way. Everybody knows that it largely grew up in consequence of the trading activities of a maritime and industrial people; and for the rest, what it acquired in war was mainly acquired as the result of our wars of successful defence against aggression. (Nor were others losers thereby: to those collective efforts in which we joined, Holland and Belgium owe their existence; so also, it is chastening to think, did Prussia, while Germany as a whole was delivered from the yoke of Napoleon.) Lastly, the facts of contemporary British policy towards the Empire are in themselves a sufficient answer to these unfounded criticisms: the whole direction of our effort for a century now has been towards conferring self-government upon our dominions overseas, and extending the principle of trusteeship in relation to native peoples.

There are those, mainly political defeatists like Bertrand Russell, who think it would have been better if we had never made the efforts we have done to secure the liberty of the lesser Powers of Europe against the aggressor, and instead had permitted Europe to be united under Louis XIV or Napoleon. This is unhistorical to the point of being Utopian. It simply

The End of an Epoch

could not have been done, not even under Napoleon, who came nearest to achieving it, as any study of the period shows. Whatever may be the utilitarian arguments for a Europe controlled by a single strong State, the force of nationality is so great that the peoples of Europe would not tolerate it. Napoleon's pursuit of European despotism was, as French observers like Talleyrand and Chateaubriand saw, a chimera: no sooner had he got Western Europe organised into the continental system than Spain revolted; then Prussia and Austria, and then the whole of Germany turned against him: for them it was the War of Liberation. That was the nearest that Europe has got to that form of unification; the moral is that that model is not likely to be successful. There will always be revolts in Europe against any one Power that tries to exert a despotism over the rest. And that is likely to be as true for Germany in the twentieth century, perhaps also (who can say?) for Russia in the twenty-first, as it was for France in the early nineteenth and for ourselves from 1763 to 1776. Happily for us, this country, except for that brief period, has never been strong enough to attempt anything of the sort. That fact has enforced upon us moderation as to ends and cooperation with others as to means, and in that has lain our security.

The whole lesson of modern European history is this: that no single Great Power is strong enough to dominate all the rest, and it really is not worth its trying. Even if it is successful for a time, the lesser Powers that have been subjugated will only seize the first opportunity of revolting, and the decision will in all probability be reversed at the next war. But we cannot go on in Europe having wars at intervals of every twenty years or so, as in the eighteenth century — modern war is far too totalitarian and destructive. The moral is, then, that we have to achieve some functioning international order, in fact a United States of Europe.

The traditional policy of this country is in no sense an obstacle to its achievement, in fact it points directly towards it; that end is its logical conclusion. It is simply not true that the past record of British policy has been to keep going an unprincipled and selfish balance of power in Europe, but

The Tradition of British Policy

rather to secure our own liberty along with that of the bulk of other European countries, whose interest is the same as ours. It is providential for us that our interest is one with the interest of Europe. That there is a widespread sense in Europe of the identity of our interest with theirs, that we are fighting their battle along with our own, may be gathered from the support that there is for Great Britain's stand and the sympathy that exists among the great majority of people in most European countries. And that has been so in all the past six years of struggle against Nazi aggression.

It must now be our business not only to carry on the war against Nazi aggression, but to follow up the lines of our traditional policy, which has been so successful in the past, by organising the opinion of free Europe in favour of an effective federation of Europe as the proper objective of the war. It is the only way out for us all, the only real solution for Europe. The League of Nations, such as it was, was the best and most hopeful thing that came out of the last war. This is not the place to discuss where and why it fell short of our hopes. But it may be suggested that perhaps it attempted too much in extending across the world from China to Abyssinia and Peru. Since the root of our troubles is in Europe, it may be well to begin with the task of building a European federation, a United States of Europe — a less ambitious aim, in the first instance, upon a surer foundation. It will be sufficiently difficult in itself, but, if we make it our objective, it should not be impossible. There could not be a more fitting culmination to the centuries of success this country has had in piloting its course than in leading Europe to this so long-desired and so necessary goal.

IV
REFLECTIONS ON THE EUROPEAN SITUATION
[Spring, 1938]

THE fundamental fact about the European situation is that the upper classes have practically everywhere regained the control which they lost by the lunacy of 1914-18. Perhaps the circumstances of those years are not exactly to be described as "lunacy", for Marxists and historians very well understand that they flowed naturally from the pre-war international order: they were the consequences of the nation-State system in the era of economic imperialism — in other words, the "international anarchy". From that, what else was to be expected? Well, now the upper classes are back again in the saddle; and, naturally enough, they bring back with them the old international order, or rather disorder. We are back in the pre-war days — pre-war in an all too threatening sense. That is the real meaning of the diplomatic tension that is going on in Europe, the relapse into the old system of alliances, under the crumbling façade of collective security, and in spite of the professions made, particularly in Great Britain, to the contrary. Here is the real explanation of the lining-up of the Powers, the swaying and changing uncertainty of the smaller States, the frantic manœuvrings for position. Position for what? — one may ask. Well, we all know, or should know, what these things are symptoms of, and to what they lead.

It must never be forgotten that this country led the way in this direction, the triumph of the European Reaction under which we live, defeated, disillusioned, driven almost to despair, like the men of 1848 living on into the age of Napoleon III and Bismarck. The year 1931 is the decisive turning-point in the history of post-war Europe. The years of the Labour Government of 1929-31 gave us the last, belated attempt to make the

Liberal post-war international order function, to get appeasement as a preliminary to disarmament and a real collective system at work in Europe. The upper and middle classes of Great Britain deliberately chose the way of reaction in that year.

1931 gave the signal to the upper classes for the recovery of their position, the resurgence of reaction, throughout Europe. Léon Blum himself once told me of the difficulties the formation of the National Government, with MacDonald at its head, created for the French Socialist Party in the elections of 1932. Before that, they had a very good chance of coming back the largest party in the Chamber, and of forming a Government. The National Government, MacDonald's defection, the " example " of Great Britain, were everywhere used with great effect against Blum and his party. The Radicals were returned much the largest party in the State, their conservative wing the stronger. That led to Doumergue and Laval, the unrestrained rule of the capitalist classes at home, deflation and depression in economic affairs, and the betrayal of the collective system to Mussolini by the French upper classes. After Hitler's arrival to power, perhaps nothing could have been done; by 1936, when Blum came in as the head of the Front Populaire, it was almost certainly too late.

In 1933 there followed the worst blow of all for Europe, when Germany went finally and wholly over to the Reaction, encouraged by important sections of the capitalist classes in this country, notably the City. In 1934 Socialist Vienna fell; that was the real end of Austria. In 1935 Mussolini won his Abyssinian empire, and the disintegration of the collective system was complete. Each stage in its disintegration, from the betrayal of the Disarmament Conference and Japan's aggression in Manchuria to Mussolini's conquest of an empire athwart our Mediterranean, African, and Red Sea routes, was accompanied by applause from important sections of English upper-class opinion.

And yet, the Reaction had not wholly triumphed. A Popular Front of Liberals, Socialists, and Communists had been called into being in both France and Spain to resist the onward march of Fascism; and for a time began to make way. But not for long. The Spanish situation revealed the underlying facts of politics at their crudest. The Spanish upper classes, aided and abetted by European Fascism everywhere — with arms and aeroplanes, recruits and technicians from Hitler's Germany and Mussolini's Italy, but supported by the active sympathies of the French and English upper classes too — entered into a rebellion against the Government elected by the Spanish people.

In France, as in England and America, the great " democratic " countries, the Right resorted to less crude methods than overt rebellion or clubbing their opponents on the head in the streets or in concentration camps. Their technique has been more elegant, but hardly less effective. In England, where the methods of the upper classes in politics are most accomplished, they found that taking in their opponents paid best: the MacDonalds, the Snowdens, J. H. Thomas. But the method they have resorted to in general in all these countries is the same: that of political, economic, and especially financial sabotage. The importance of this technique, and particularly the regularity and generality of its use, has not been much commented upon as yet, perhaps not even widely noticed. Yet it was extremely effective in this country in the last months of the Labour Government: the constant denigration and undermining of confidence in Britain's financial position in the capitalist press, the daily lucubrations of the City editor of *The Times*, the persistent suggestion of " crisis ".

Since then the technique of sabotage has become much bolder, more unscrupulous, more shameless. France, where the Left is stronger than in most countries, has been the chief field for its exercise. A Left Government in France is always met by a run on the exchange; and if this is not sufficient to bring it down, as with Herriot's Government of 1924 or Blum's in 1937, there is always the Senate, like the House of Lords, in the background, ready to do its duty. It was the Senate that wrecked Blum's second Government this spring. " *Moi, j'ai confiance dans le Sénat* ", said the middle-class proprietress of a bookshop to one of her bourgeois customers in my hearing at

Reflections on the European Situation

Avignon, the morning that Blum's financial plan was published. The Senate justified her confidence and that of her class.

There are all sorts of morals pointing in every direction, any number of lessons in practical politics, to be drawn from that by the Left. It looks in this year of grace 1938 as if it is now impossible for any Government of the Left to govern against the will of the capitalist classes. Even Roosevelt's administration, with an enormous majority in the country behind it, has been obstructed, stultified, and for long periods rendered almost null and void by the sabotage of the Right. They just will not play. Perhaps one ought not to be surprised; it is what one should expect. But the lengths they will go to is surprising; one cannot help being morally shocked, when one remembers that the whole appeal of the Right is to patriotism and national feeling, that many elements of the Right do not mind the ruin of their country, provided that they retain their position of privilege and power over their compatriots among the ruins.

The behaviour of the Right in France has been beyond belief shocking. With patriotism on their lips and all the time vilifying Blum as a Jew and an internationalist, they have seized any and every occasion to bleed the country of capital, export their holdings, make a profit on each devaluation which they have rendered necessary by their financial sabotage, and salt the proceeds safely away abroad before the next. They have been determined by hook or by crook to bring down the Front Populaire Government resting on a large popular majority in the country. At a time of immense national danger, and in face of the national enemy, they have been prepared to weaken France, refusing to allow a real national Government to be constructed on the basis of the popular majority, rejecting Blum's patient and patriotic offers with contempt and contumely. No one can deny that Blum has acted with a great sense of responsibility and loyalty to the interests of France — indeed critics on the extreme Left may say that in some respects, notably with regard to Spain, he has gone too far to placate the enemies of the working class for the sake of national unity in confronting the European danger. No one can deny that Blum is a man of integrity, singularly incorruptible in the miasma of French politics. A proper investigation of the *bas-fonds* of the Right in France would reveal much that is not only unsavoury, corrupt, but a good deal that is actual treason against the State. There are groups on the Right that are in touch with Hitler and prepared to do his bidding; there are others who take their orders from Mussolini — it seems that the Roselli brothers were murdered not by Italians, but by Cagoulards acting on instructions from Rome. While Dormoy was at the Ministry of the Interior these investigations were beginning to be made; that was another reason why it was so essential that Socialist government should come to an end. No one on the Right wanted the chances of an understanding with Mussolini prejudiced.

It is interesting to observe the persistence with which they go on hoping to pick up the crumbs from the table at which Mussolini and Hitler feed together; though what France can conceivably gain from an agreement with Italy in which France makes all the sacrifices, it is hard to see. Do they think that they can get Mussolini to fight Hitler for them by making concessions to him, giving up Spain to him for example? What a fearfully dangerous game for France they are playing. Italian policy has been a *politique de pourboire* ever since 1860. If Mussolini gets concessions out of France (the same holds good of Great Britain), he will only use them to get a higher price out of Hitler in the next round — perhaps Corsica, Nizza, Tunisia — who knows? The only safe way for France, as for England too, is to see that the Spanish Republic holds on so long that Mussolini and Italian Fascism are exhausted and over-strained by the double effort in Abyssinia and Spain. Then there may come about the sort of régime in Italy with which a defensive bloc of Western Powers is possible — Great Britain, France, Italy, Spain: it would be a strong bastion of peace in the West and in the Mediterranean. Moreover, we should not then be so powerless to affect the issue in Central Europe. Along with Czechoslovakia and Russia, we should be in a strong enough position to maintain peace against would-be aggressors. This is the conception of foreign policy that not

Reflections on the European Situation

only the Left in Great Britain and France stands for, together with the great bulk of the smaller Powers and the democratic forces everywhere; it is also the conception of policy that the vision and understanding of Mr. Churchill sees to offer the only security for the safety of this country and the peace of Europe.

It has been the happy good fortune of this country, and the sheet-anchor of our security throughout our history, that our interests have coincided with those of the great majority of European countries as against those of any too-powerful State whose aggressive designs have threatened the independence and sometimes the very existence of the rest. This traditional policy, which certainly served our ends, also served the interests of the rest of Europe on the whole, or they would not have adhered to it. The League of Nations was an attempt to generalise this conception and enable it to serve world-purposes. That is why British policy is peculiarly closely associated with the League idea. That is why all those elements in Great Britain that have so disliked and undermined the League have done such a disservice to British interests, no less than the interests of Europe as a whole. Hence, too, the reason for the failure, the confusion, the contempt into which British policy has fallen, the disintegration of the League and of the European system, since 1931 — since the British governing classes have been distracted between the contrary pulls of their class-sympathies abroad and what is the plain interest of this country and of Europe as a whole.

It is the sabotage of large elements among the upper classes in England and France — for Hitler and Mussolini have many friends among both — that is responsible for the frustration and confusion in the foreign policy of the democratic countries. And it can only be ended in one way — by the Left gaining power and making it operative. It is all a problem of power, and it is a question whether the Left in these years and in these countries has ever yet attained sufficient power to make its will prevail. And it must be admitted that, apart from its regrettably insufficient sense of power, its lack of realism and determination, the Left plays into its opponents' hands in the matter of financial and economic sabotage. It makes the mistake of demanding more economic concessions than the existing system can stand; when the breakdown comes, therefore, in increased unemployment, falling-off of production, rising costs and prices, the Left in power, or rather in office without real power, is made to take the responsibility. There was a certain element of this in the "crisis" of 1931. But the introduction of the forty-hour week in France is a glaring example. Almost everyone of any intelligence on the Left would now admit that it was a mistake; that it was trying to bite off much too much at one time, that it has imposed an altogether too heavy burden upon the French productive system. It is not fair to blame Blum for it, however: it was the result of the spontaneous pressure of the working classes after the long strain imposed upon them by the deflationary measures of Laval, issuing in the series of successful strikes which accompanied the beginning of Blum's Government. It may be opined that Blum himself knew it was a mistake and that the masses had gone too far. But here, too, there lies a moral of the greatest importance for the Left. The masses must be given leadership, and strong leadership. It is no good leaving them to their own devices: they do not know what is good for them. It is disastrous for the leadership to follow in the wake of the rank and file. It is humiliating and dangerous for a leader of the working class to have to go down like Blum and *plead* with the workers on the Paris Exhibition to work hard enough to get it finished. They must be made to work, in their own interests, in the interests of their class and their country, and in order that their cause may ultimately triumph.

Of course, there were extenuating circumstances. But it would have been far better if Blum had strengthened his hold on power and stayed there. When once you have full control of power you can make what reforms and concessions you choose later on as opportunity offers. It is *power* that counts in politics, not momentarily pleasing the people.

If this holds good in the sphere of social reform, it is still more true in that of high financial policy. This is the rock

Reflections on the European Situation

upon which social-democratic Governments are apt to split almost everywhere. The financial policy of the first Blum Government was characterised by a deplorable spirit of indecision, hesitation, and compromise. Blum himself can hardly be held responsible for it. He was faced with a succession of crises in industry, which he handled with skill and on the whole with success, a perpetual crisis in the foreign field which the Spanish Civil War threatened to turn into a European War. It does not seem that Vincent-Auriol, Blum's second-in-command, had the grasp necessary for the Ministry of Finance; to some extent the situation resembled that in which the Labour Government was placed through the incompetent obstinacy of Snowden in 1930-31. A succession of mistakes was made. Devaluation ought to have been effected straight away; instead of which the Government waited until millions of gold francs had been lost to the reserve, and a vast flight of capital abroad had taken place. But far more important here, again, than any actual financial measures, was the question of power. The Government should have assumed complete control of the financial position, with exchange control and all, from the first. There would have been the opposition of the Radicals, or some of them, to consider. But they would not have dared to break the Front Populaire in all the strength and enthusiasm in which it then stood with the masses.

The moral for any Government of the Left is that it must always make the utmost of every opportunity that is presented; it must exploit its initial advantage to the very maximum. The Labour Government when it came in with a good deal of enthusiasm behind it in 1929 did no better, indeed a great deal less well, than Blum in 1936-7. But though he gave the French workers the five-day week and a new industrial charter, he failed to entrench himself in power. When he came back to office in March of this year, he had learnt, intelligent man that he is, the lessons of his first administration. He took over the Ministry of Finance himself, and prepared a financial plan which for the first time covered the whole French financial situation and made proposals for dealing with it as a whole.[1] The plan was masterly; but by now the political opportunity for applying it had gone by. The enthusiasm and strength behind the Front Populaire had greatly subsided; the French bourgeoisie were no longer afraid; the Senate treated him, the Prime Minister of France and the most distinguished person in French politics, with open contempt. Next day they wrecked his Government, and with impunity. What morals there are for the Left to learn — if only they are capable of learning them! The chief of them all are: (1) to go slow as regards economic reform until you are sure of *power*; (2) always exploit what opportunity is given you right up to the maximum.

It is much the same story with the Spanish situation and the policy of non-intervention. Blum has been greatly blamed by the Left for instituting it. But we do not know the full circumstances of the case as yet. It is fairly clear that the responsibility is not his — except in the sense that if he had been a Hitler or a Mussolini, *i.e.* a real leader and not the head of a parliamentary coalition, he would have seen to it that his will prevailed. There was the British Government with its clear indication to the French that it would go no further than a policy of non-intervention. There was the extremely dangerous internal situation in France, with the army largely officered by Catholics: open intervention, the only proper reply to Hitler and Mussolini, might have meant civil war in France, as they well knew. There was probably nothing for it but non-intervention for Blum. And non-intervention on the assumption that it would be adhered to, or at least not too widely departed from, was probably the best policy all round. Nobody can have foreseen the extent to which the Fascist Powers would break their agreements — the actual waging of a war upon the Spanish people. Certainly the Spanish Government itself had no conception of the lengths to which Mussolini and Hitler would go. A member of the Labour Party delegation sent to Paris in September 1936 to consult

---

[1] Cf. the very favourable opinion of *The Times* economic correspondent, which Blum quoted at length in his speech to the Senate, *le Populaire*, April 13, 1938.

Reflections on the European Situation

Blum told me that the Spanish Government had assured them that, given reasonable non-intervention, they would win the war. Perhaps one more example of the ridiculous optimism of the Left in politics.

But surely no one should have trusted Mussolini's or Hitler's word for a moment; they are the self-declared breakers of agreements; if it happens to suit your interest to break a pledge, then break it — that is their creed. It has served them well, so long as fools allow themselves to be taken in. *The non-intervention policy should have been made conditional from the first* — like the subsequent Anglo-Italian agreement. A conditional non-intervention agreement might have had the effect of very considerably limiting the aid Franco received from outside. If there were any difficulty in the matter, the French and ourselves could have used our overwhelming naval power to put a cordon round Spain and seen to it that there was genuine non-intervention.

It is the Machiavellianism of the Fascist Powers that has changed all that. And the only way to meet Machiavellianism is with Machiavellianism. In the latest phase of the Spanish conflict it appears that the French have realised that, and are prepared to play Mussolini and Hitler at their own game. But what makes it so difficult to meet them on their own ground is the sabotage of large elements among the upper classes.[1] Hitler and Mussolini know that they can count on their aid to distract and confuse the will of the " democratic " countries. It is the European class-conflict cutting across the former national divisions that complicates the picture. There is nothing more disingenuous and insincere than the parrot-like reiteration of the British Government that they will have no part nor lot in the ideological conflict that is dividing

---

[1] Cf. the revealing, if restrained, words of the Paris correspondent of *The Times* on the Czechoslovak crisis (May 23, 1938) : " In Berlin, it is felt, there can be no illusions about the position of this country, provided that Herr Hitler forms his judgment upon the communications received from his duly accredited diplomatic representatives. It is known, however, that reports on French conditions reach him through other agencies, which are often in touch with elements less representative of French foreign policy. The dangerous potentialities of such unofficial diplomacy are only too evident."

Europe. In fact they have played a most important part in it, and it is clear on which side. There is no doubt that the British Government, no less than the upper classes, want Franco to win in Spain — though it plainly goes against the interests of our country.

One may go further: it is British policy in these years since 1931 that may be held responsible for the disintegration of Europe. By our position in Europe and outside we are in a place to lead the European Concert. The point was made to me last year by a very penetrating Italian publicist that the disorganisation of contemporary Europe is due to our abdication. No doubt Mussolini regards it as our degeneration, and hopes to take our place in the Mediterranean, if no further afield. The clue to that abdication, the loss of our former leadership in Europe, is to be found in this — that since 1931 the British governing class, or large sections of it, have been following their class-interest in the conflicts raging abroad rather than what is to our national interest. At the very least they have been confused in purpose by the dichotomy; at the worst they have been guilty of collusion. Hitherto in our history the problem has not much arisen, for the two, the class-interest of our ruling classes and our national interest, have coincided. But with the challenge of the working-class movement to their hold on power, the situation has changed. One cannot avoid seeing that the one line that runs consistently through the tergiversations, the betrayals, the confusions of British policy in these years is the determination not to see a victory for the Left anywhere, certainly not to aid it — even though it is to the interest of this country that we should. What a change from the days of our great success as a nation, from Canning and Palmerston! No wonder it has brought us to such humiliations, to such a pass as we are in now.

Take the cardinal importance of our anti-Russian policy in disorganising Europe and the Far East. British policy has played into the hands of the Communist leaders in keeping Soviet Russia isolated from normal and helpful contacts with the West. And this, even though it is our plain interest to collaborate with Russia against aggression in the Far East

Reflections on the European Situation

and in Europe. The result is that Japan has been enabled to run amok over China, conquer large tracts of the country, destroy the peace and threaten our large interests there. It has been the consistent aim of British imperialism to weaken Russia *vis-à-vis* Japan. Now that they see the danger from Japan, they are willing enough to collaborate with Russia in aiding the Chinese. Practically all sections of the British upper classes would now welcome a Japanese defeat. How ironical it all is! For they are largely responsible for the over-development of Japanese power, as they are for Mussolini's in the Mediterranean. Both could have been dealt with quite simply, and without much danger to ourselves, if our ruling classes had played a straight game with collective security since 1931, in friendly association with the U.S.A. on one side and Soviet Russia on the other. In fact, however, throughout the Far Eastern crisis of 1931-2 practically the whole of the Conservative press and the Conservative Party were on the side of Japan — at the risk of alienating America, at the risk of raising up a first-class danger to ourselves in the Pacific.

The same is true of the situation in Europe. For there can be no overwhelmingly strong system of collective security against the aggressor Powers in Europe, unless Russia is in it. How clearly Mr. Winston Churchill sees this, and yet how timidly he has to put the point in his campaign speeches for collective security, because of the susceptibilities of Conservative opinion. They *will* not take the necessary steps to safeguard our security and preserve the peace of Europe, because of their hatred of Communism. They are first-class material for the psychological appeal of Hitler's disingenuous campaign against Bolshevism, designed to split the " democratic " countries in two and frustrate their unity and effectiveness in action. The only alternative policy that the governing class can put forward is the desperate gamble of making concessions to Mussolini and Hitler, in the name of conciliation and with the hope of buying them off. It is the policy of paying Danegeld.

We have seen that there is no reason to suppose that, however much they concede to Mussolini, he can now move

The End of an Epoch

away from the Berlin axis and come over to the side of Great Britain and France; or, secondly, that there is any limit to his ambitions in the Mediterranean. Nor, again, is there any more hope of success to be gained from the second objective of their policy, buying off Hitler: " conciliation " with Germany, as they put it. For that involves accepting Germany as the dominant Power in Central Europe, if they *will* exclude Russia. It means a German Mittel-Europa, and the return of the ex-German colonies too. All that we fought the war of 1914–18 to gain will have been lost by the supineness, the fatal confusion in aims, the indifference of the British governing class since 1931. They had it in their hands to save the peace of Europe; they had only to follow our proud, traditional role of organising and leading the majority of European Powers — the core of a collective peace system. But they refused to do it; some of them have done all they could to advance the cause and clear the way for the aggressors. Their responsibility is clear when one reflects that no one foresaw a war on the European horizon in 1931; whereas, now, no one can exclude its possibility.

But it looks at last as if the facts of the European situation are driving the British governing class to a clearer alignment, now that the confusions of the past seven years have brought us into imminent danger of war. Though the British Government is mainly responsible for selling the Spanish Republic, in the hope of buying Mussolini's support, they seem at last to have realised the danger of allowing Czechoslovakia to be broken up and absorbed by Germany.[1] The importance of Czechoslovakia is that it is a microcosm of the whole European situation. If its independence goes, it will mean that we are faced with a Europe under German domination. It may be

---

[1] Even this expectation proved to be too sanguine, so utterly blind and senseless was the policy pursued by the Chamberlain Government. No one could have foreseen that, so far from joining with Russia and France to safeguard the independence of Czechoslovakia, Chamberlain preferred to join with Hitler and Mussolini in ending it. The refusal to reach any agreement with Soviet Russia in time was the key to the disintegration of Europe. It gave Hitler his opportunity and made the war inevitable [1945]. Cf. in detail, L. B. Namier, " Coloured Books ", in *Political Quarterly*, April–June 1945.

Reflections on the European Situation

that at the last moment the British Government will join with France and Russia in safeguarding the independence of Czechoslovakia. That would be the better way; but what nonsense it makes of British policy in the past seven years; what bankruptcy in retrospect it reveals!

The issue in Europe seems to be clarifying itself along these lines, though British policy will do all it can to avoid a clear decision. For a clear decision might aid the democratic Powers, the forces of the Left throughout Europe. It might bring about the end of the Reaction: it will not last for ever. The British governing class well understands, none better, how precarious is the hold of the Reaction even in the moment of its widest success. It might be defeated or checked somewhere or other — in Spain, or in Italy (1935!) — and then what would follow? Hence the British Government's insistence upon a policy of concessions to Fascism, "conciliation" all round — at the expense, of course, of republican Spain and democratic Czechoslovakia. They are instinctively right — if not the longest-sighted governing class in Europe: they know that another war would destroy their order, and they are out to preserve it. Whether they will succeed over the next phase in Europe's history or no, depends on whether Nazi Germany can play the game of moderation, content to fulfil her aims gradually, piecemeal, by economic expansion in Central Europe rather than by military aggression.

Royer-Collard said of the July monarchy that Louis-Philippe saw very well that France could not stand a war; what he did not see was that France could not stand a peace. It may be the same with Hitler. If the Nazi régime can pursue its aims moderately, without bringing about a war, then there will be peace in Europe — for a time. Nobody supposes that it can last beyond a certain time, so long as the present international order remains what it is. Meanwhile, so long as there is not a war, what a good thing the Nazi régime is for the upper classes of Europe. Not only in Germany, where they have been "saved" from Liberalism and Socialism, but everywhere else: the danger from Nazi Germany keeps the Left in France and Great Britain in order, stultifies their social aims, brings them into line over national defence, prevents them from gaining power, or if they gain it, from exercising it. Really, if Hitler did not exist, it would be necessary to invent him; and some of the upper classes have had a pretty good hand in inventing him and keeping him going. But the logical end of the process, if some measure of collective security is not re-established, is a Europe dominated by Germany.

I have long been haunted by the thought that the upshot of British post-war policy, its evasions and confusions and duplicity, might be to postpone the day of reckoning for a time, at the expense of making Germany ever stronger against the rest of Europe. So that when our generation comes to face the responsibilities bequeathed to us, we may find Germany so strong that we cannot resist and are faced with defeat for this country and our friends later in our lives. It is by no means an impossible prospect. When all would have been so much easier if only our ruling class had played straight with the collective system since 1931. But on the basis of the existing order no collective system is ultimately possible. The conflicts of national State-sovereignty are ineluctable; a true functioning international order presupposes the victory of the Left, at any rate in Great Britain, France, and (probably) Germany. Meanwhile, it is the nature of the existing system to keep us ever on the brink of war. And, as I say to the "democratic" electorate in my own constituency: if you like it, go on voting for it. Alas, they do not understand what one means; they go on voting for it; and the tragedy is that they will pay for their folly not in their own lives — that would only be justice — but in those of their sons.

V

THE END OF AN EPOCH

[June 1940]

"To tell the truth needs no art at all, and that is why I always believe in it."—Mr. STANLEY BALDWIN on "Rhetoric", Cambridge Union, March 1924.

"I have just assumed the duties and responsibilities that attach to the office of Prime Minister. I am entering on them at an age when most people are thinking of retiring from active work, but I have hitherto led a sober and a temperate life. I am informed that I am sound in mind and limb, and I am not afraid of the physical labours which may be entailed upon me."—Mr. NEVILLE CHAMBERLAIN, speech on being elected Leader of the Conservative Party, May 31, 1937.

A PERIOD of nine years, one of the most fateful in the history of our country, came to an end with the overthrow of the Chamberlain Government. It is probable that future historians will be hard put to it to find another period comparable to it in folly and disgrace, in corruption of the very sources of judgment, in lack of vision and criminal obtuseness in high places, unless they go back to the twelve-year rule of Lord North and George III that ended in the loss of America. It is impossible here to do justice to the humiliating period that we have lived through since 1931 : it would need the pen (and the space) of a Swift or a Voltaire, the brush of a Hogarth or a Rouault, the pencil of a Low.

What are we to think of a period of which the decisive political manifestations were the two trick elections of 1931 and 1935 — the former based on financial panic, the latter on a deliberate fraud ; the solid mass of Tory yes-men in the House of Commons who sat there year after year voting for the wrong things, backing the wrong horses on the wrong courses, greeting the loss of strategic position after position to our enemies, Manchuria, Abyssinia, Spain, Austria, Czechoslovakia, if not

The End of an Epoch

with rapture, at any rate with complacency and cheers; yes-men who would listen to no warnings from Mr. Churchill, Mr. Eden, Mr. Amery, Lord Cranborne, Mr. Duff-Cooper, let alone Lloyd George and Liberals and Labour men; yes-men whom nothing would unsettle, so tenacious was their misguided instinct of self-preservation, until the military disasters to their country blasted them out of their position after those two days' momentous debate? What are we to think of a period whose typical diplomatic representatives were Sir Francis Lindley at Tokyo, Lord Perth in Rome, Sir Nevile Henderson in Berlin,[1] with Sir John Simon and Sir Samuel Hoare in charge of the Foreign Office? Whose great statesmen, responsible for the lives of forty million Englishmen and the safety of a great Empire, were Mr. Baldwin and Mr. Chamberlain? The head of a family with a great tradition in our history has said: "It has been my ill-fortune to have lived under the two worst Prime Ministers that England has ever had". That is perhaps going a little too far, but not very much too far.

It is only possible here to single out one or two of the leading characteristics of this period now at an end, to diagnose its peculiar quality, perhaps rather its pathology.

And first it should be said that this period is really a continuation, and a worsening, of the years since 1922. It is in the past nine years that we have seen the fruition of the mistakes, both of commission and omission, of those earlier years. Yet the situation might have been saved after 1931: all was not yet lost; but, instead, we have watched, with the fatal fascination of some tragedy being enacted upon the stage of Europe, every opportunity being lost, every wrong turning taken, the whole thing muddled and mishandled. How has it

[1] We have been favoured with Sir Nevile Henderson's own account of his reactions on his appointment (*v.* his *Failure of a Mission*, p. 13): "In the first place, a sense of my own inadequacy for what was obviously the most difficult and most important post in the whole of the diplomatic service". In that, at least, he was evidently right enough. "Secondly, and deriving from the first, that it could only mean that I had been specially selected by Providence for the definite mission of, as I trusted, helping to preserve the peace of the world." That in itself should be enough to disqualify him from such a post: a completely mistaken conception of an ambassador's function.

The End of an Epoch

been possible? After the last war, the great democracies took the opportunity to throw over the first-class men who had led them to victory, in favour of comfortable second and third raters. The United States threw over Woodrow Wilson for the miserable Harding and the unspeakable Coolidge; France threw over Clemenceau; in Great Britain, the Conservative Party threw out Lloyd George for a Bonar Law and then Baldwin and then Neville Chamberlain. During all these years from 1922 to 1940 the Conservative Party was responsible for keeping out the one man of transcendent political ability this country had, at the height of his experience and European prestige.

But the period since 1931 has had its own peculiarly nauseating quality. Nor is it difficult to define. Apart from the disasters that have overtaken our policy, the failures that all can now see, and which it is not my purpose to go into, there has been a lack of candour, government by fraud and deceit, that has sapped and undermined the old efficacy of parliamentary and democratic institutions in this country. Mr. Baldwin said in his Rectorial address to the students of Edinburgh University on " Truth and Politics ":

Though it has been accepted through the ages that half a loaf is better than no bread, half a truth is not only not better than no truth, it is worse than many lies, and the slave of lies and half-truths is ignorance. Ignorance, static and inert, is bad, but ignorance in motion, as Goethe once observed, is the most terrible force in nature, for it may destroy in its passage the accumulated mental and material capital of generations.[1]

Lord Baldwin should know. He is no doubt right: the half-truths that were ladled out to an ignorant electorate by the National Government, from its very inception to its end, were in the event more disastrous even than the many lies of Hitler and Goebbels. For they, at any rate, never disguised from their people the desperately serious nature of the struggle upon which they had embarked them, and the heroic efforts that were required from the German people in the course of it. Whereas Mr. Baldwin, who knew for two whole

[1] Earl Baldwin, *On England*, pp. 96-7.

The End of an Epoch

years previously the danger into which our relations with Germany had passed, confessed in his swan-song as Prime Minister, that he had not told the people, for he " could not imagine anything that would make the loss of the election, from my point of view, more certain ". After that, he ceased to be Prime Minister and was succeeded by Mr. Chamberlain. But that did not improve matters.

The fact was that the real reason for our demoralisation as a people went far deeper than the shortcomings of these two men; it was to be found in the very nature of the National Government. That Government had the lie in the soul from the very beginning. It was founded in fraud and it lived by fraud. Its purpose in 1931 and onwards was to collect together all the electoral support it could from all quarters by all means in order to isolate the Labour Movement and keep it out of power for good. The very idea of National Government was thus to undermine and destroy the effective working of parliamentary democracy, in which Government was kept up to the mark by a strong Opposition that could be called upon to replace it if things went wrong. It brought into being a coalition of all classes against the working class. It does not affect the analysis in the least that the majority of people voted for them. The old rationalist assumptions of liberal democracy were out of date, even if they ever did apply. Conservatives well knew that. Actually the real interests of 75 per cent of the community were represented by the Labour Movement, even though they may not have realised it,[1] and the patriotic interest of the country's security as well, though the Labour Party's spokesmen were incapable of putting it across.

In these circumstances, the technique of " National " Government was to soothe and swaddle opinion, to let no clear ideas of the issues at home and abroad get hold of the minds of the people, to lull them into a sense of false security when the gravest dangers were accumulating for our country abroad. It must be said that the responsible Ministers and their henchmen in the press were great masters of the technique.

[1] Cf. Rousseau: " What makes the will general is less the number of voters than the common interest uniting them ".

The End of an Epoch

They ruled the public mind by suggestion. Anyone who wished to rouse the public to the dangers impending was an " alarmist ", or, to quote the elegant word of that great statesman, Sir Samuel Hoare, a " jitter-bug ", if not a " war-monger " ; those who saw all that we stood to gain by upholding collective security in Europe were denounced by Mr. Chamberlain himself as " blood-thirsty pacifists ". The collective security system of the League was to him " mere midsummer madness " ; and selling it to our enemies was " realism ". (Perhaps " selling " is a misnomer, for they were not even paid for the services they rendered.) All was for the best in the best of all possible worlds. Was not Mr. Baldwin Prime Minister, and after him Mr. Chamberlain ? Was not Sir John Simon at the Foreign Office and afterwards at the Exchequer ?

It was extraordinarily effective. It was impossible to get any sense of the real situation across to the people. They were lulled to sleep, they were doped and duped until they woke up to find themselves engaged in a war for their very existence, and inadequately prepared for it. Bitterly as one may feel about the matter as a political candidate responsible for one's own constituency, it is as nothing when one thinks that all the warnings of Mr. Churchill, with his immense experience and prestige, his oratory and command of the pen to recommend him, went as unheeded as those of any Labour candidate. You have only to read his collection of speeches, *Arms and the Covenant* — or Mr. Attlee's, for that matter [1] — and compare them with Mr. Chamberlain's, to gauge the extent of the tragedy, the national folly.

Yet nothing could be done about it until, after nine months of war, nine months of defeat and disaster, of blundering and indecision, we had reached the very verge of the abyss. The extraordinary thing was that the more these men failed, the more they held on. There can never have been such a religion of failure, such tenacity and faith in being wrong. The more Chamberlainism failed, the more disasters it had to its credit, the more people clung to it and its visible expression, the late

[1] *v.* C. R. Attlee, *War Comes to Britain.*

The End of an Epoch

Prime Minister. I remember now an idiot woman in my own constituency — she was an average woman of the Conservative Party, a member of her Women's Institute — who said to me after Munich, with that fanatic look in her eyes: " Mr. Rowse, don't you think that Mr. Chamberlain is the most wonderful man who ever lived?" A friend of mine, only the other day, before the overturn of the Chamberlain Government, was in converse with a banker in the train, who told him that this war was very like the last, except that in this one, under Mr. Chamberlain, we had so much better leadership. Poor Lloyd George!

Whatever can explain such a state of mind? How can it be accounted for? Mr. Harold Nicolson has made some brilliant observations on the fascinating, if repulsive, phenomenon of Chamberlainism:

> The true nature of the devotion which he inspires in so many of his followers is something which ought to be analysed and examined by those who are interested in the condition and future of the Conservative party. It is not the descendants of the old governing classes who display the greatest enthusiasm for their leader; it is rather the descendants of the industrial revolution. Mr. Chamberlain is the idol of the business men; they feel that he understands their perplexities, their ignorance, their sad little optimisms, their harmless ambitions; they have during all these years identified themselves with this representative of the backbone of England, and any decline in his prestige is a decline in their own.[1]

Very illuminating, very true. The fact is that Chamberlainism is a phenomenon of decadence, of decline: it is an expression of the declining hold of the rich middle class, who did great things for England and for themselves in the nineteenth century, upon the life of the country. After the last war, but especially after their experience of the world depression which coincided with the Labour Government of 1929–31, they felt themselves to be on the defensive. They were determined to hold on to what they had got, a sentiment which they successfully communicated to a vast number of people whose little property was in no danger from a

[1] *The Spectator*, May 17, 1940.

The End of an Epoch

Labour Government. But it has been the property-sense that has dominated the years since 1931, until the fact became too evident that you could not conduct a European war on it. That was at the bottom of all their politics from 1931 onward. In the old days, the Conservative Party had principles which it was possible to defend, and to respect if you did not agree with them. Men like Disraeli, or Salisbury, or Balfour had an intellectual position that could be defended as against that of Gladstone, John Morley, or Asquith — or even Mill or Marx for that matter. But the National Government could in its very nature have no principle, except that of holding on to what they or their supporters had got. It was therefore always open to interest pressures, and pursued an unprincipled concessionism — making concessions to whatever interest was strong enough to bring pressure to bear, even to sections of the Labour Movement, to the Trade Unions, to the weaknesses of democratic opinion — as the price of remaining on in office.

The effect was utterly demoralising to all concerned, corrupting to political morale, and ruinous to the country. In the circumstances of the modern mass-electorate, leadership becomes a matter of infinitely greater importance than ever before; it is vital that the morale of the people should not be sapped by telling the people only what the people love to hear. There is something seriously wrong with parliamentary democracy as we have known it in the years since the last war, though what has happened since 1931 is not so much the fault of democracy itself as of the use which has been made of democracy to trick and deceive it, and govern by fraud and humbug. The National Governments could not have continued to hold power if the real nature of their hold upon power and the facts of the situation had been realised by the mass of the people. No doubt it could be argued, better fraud than force; but fraud can be as destructive as force.

Whence, it may be asked, came that peculiar flavour, the moral tone of Chamberlainism, that extraordinary smugness, that self-righteousness, that impermeable and fatuous complacency which all the chosen, inner circle exuded? It was a chief instrument in their technique, by which they muffled all clarity of thinking in the country, and stifled all criticism; it was a defence-mechanism which they all used, much as a grass-snake emits an unpleasant odour with which to defend itself: it was their characteristic smell by which you could recognise them a mile away. I offer this suggestion towards a more complete analysis of Chamberlainism when the time comes. It has not been noticed how closely connected was the inner circle of Baldwin-Chamberlainism with the higher ranks of Nonconformity. Nonconformity in the nineteenth century had a contribution, of a sort, to make to English society; social progress and humanitarian courses owed something to the Nonconformist conscience and its power of moral indignation. But political Nonconformity in deliquescence, without its conscience, has been a sorry spectacle in these past years; there remained only its peculiar brand of self-satisfaction. It was very strong with the men of 1931, the men of Munich. Mr. Baldwin boasts of his descent from a long line of Nonconformist divines; Mr. Chamberlain is a Unitarian; Sir John Simon the son of a minister; Sir Samuel Hoare descended from Quaker stock; Lord Runciman the *beau idéal* of the Wesleyans with their worship of wealth and worldly success; there is Mr. Ernest Brown, whose favourite activity, I gather, is local preaching.

Many of these characteristics are to be noted in precise and documentary form in Mr. Chamberlain's speeches: they make such illuminating, if sickening, reading now.[1] Still, the personal characteristics of Mr. Chamberlain have some historic importance; for, strange as it may seem in the case of so obviously second-rate an individual, his was peculiarly a period of personal rule. No one can say that his was not a strong personality: he stamped it upon his administration as no other recent Prime Minister has done. He pursued a *personal* foreign policy to which he sacrificed his Foreign Secretary and Under-Secretary, and the corporate, accumulated knowledge of the Foreign Office itself. When he went to Munich, he did not take the Foreign Secretary or any author-

---

[1] *The Struggle for Peace*, by the Rt. Hon. Neville Chamberlain.

The End of an Epoch

itative representative of the Foreign Office, but his *alter ego*, the Chief Industrial Adviser to the Government, Sir Horace Wilson. For this person's knowledge or competence in the realm of foreign affairs there is no more evidence than in the case of Mr. Chamberlain. The result may be seen today.

In reading through these speeches one gets the clearest impression: the vanity and self-satisfaction of the old man arrived late in a high office, his ignorance of foreign affairs, his utter lack of vision. We find him again and again defending Mussolini's " good faith " in the House of Commons. In October 1937 he was insisting against Sir Archibald Sinclair that there was no official confirmation of the landing of Italian troops in Spain; the occupation of the Balearics was an " unfounded idea "; the movement of Italian troops into Libya at that time had no " connection with present events ".[1] No wonder poor Mr. Gallacher was reduced to interjecting " Nothing means anything ": that exactly described the situation. The martyrdom of a friendly Republican Spain was allowed to go on; no notice taken of the bombing of Guernica, in which the Nazis tried out the methods subsequently employed against Poland, Norway, Holland, Belgium, France, and which, alas [June 1940], we daily wait to experience ourselves. British ships that took supplies to the Spanish Republic were stigmatised as seeking only profits, a curiously motivated attack for an ardent believer in the profit-making system; they were to enter Spanish ports at their own risk. Justice for the Spanish Government was described as " intervening on one side ".[2] The whole idea of collective security, with which our own safety was bound up, is misrepresented as " going to war with everybody ".[3] The low-down cliché of the petit-bourgeois mentality that the collective system meant being " the policeman of the world " is repeated to the herd of cheering Tories.[4] No wonder he was the idol of the smallminded business man. No conception that it was our own country's safety that was being betrayed bit by bit into the hands of our enemies even entered his mind or theirs. How different from Mr. Churchill's informed and historic percep-

[1] *Ibid.* pp. 36-40.   [2] P. 58.   [3] P. 59.   [4] P. 60.

The End of an Epoch

tion that our safety was bound up with that of the other European peoples against the Fascist aggressors.

Our security on that flank was betrayed in Spain. Yet in the New Year of 1938, " I do not want you . . . to think that I am anxious to be relieved of my office ".[1] There is all the tenacity of the septuagenarian in that: would that he had been relieved! Instead, it took a series of disasters and a political revolution to remove him, and even then, to the last, he was appealing to his " friends " in the House. Over Czechoslovakia and the deepening danger it was the same story: in July 1938 he had " never enjoyed better health and spirits "; his efforts had " already changed the atmosphere on the Continent for the better ".[2] In September he was accepting Hitler's assurance that his Sudeten demands were his last territorial claims in Europe.[3] Lord Runciman, an appropriate choice, was sent to facilitate the break-up of Czechoslovakia. We know from the notorious interview with American journalists that appeared in the *Montreal Star*, that the Prime Minister of Great Britain had committed himself to the view that neither Britain nor France nor Russia could fight for the defence of Czechoslovakia even if they wanted to, that Czechoslovakia " could not survive in its present form " and that " frontier revision might be advisable ".[4] Hitler and Ribbentrop must have found all this distinctly encouraging.

Even more important was the obvious distaste for any common policy with Soviet Russia, in that reflecting the attitude of the whole class he represented; while the exclusion of Soviet Russia from the discussions concerning the fate of Czechoslovakia was fatal not only in that particular matter but also to the development of a common front with the Western Powers. When Hitler presented his ultimatum, with demands that threatened the Republic's existence, the British public was told in a broadcast of unparalleled meanness that it was " a quarrel in a far-away country between people of whom we know nothing ".[5] The Munich Agreement was

[1] *The Struggle for Peace*, p. 65.   [2] Pp. 237, 241.   [3] P. 275.
[4] R. W. Seton-Watson, *From Munich to Danzig*, pp. 38-9.
[5] *The Struggle for Peace*, p. 275.

The End of an Epoch

described by Mr. Chamberlain in Disraeli's words as "Peace with Honour":[1] no one observed at the time that Salisbury afterwards admitted that at Berlin in 1878 we had backed the wrong horse. The Agreement was also described as bringing "Peace in our time".[2] It lasted precisely eleven months. Yet after Munich Mr. Chamberlain insisted: "I have nothing to be ashamed of. Let those who have, hang their heads."[3] In December he was "not conscious of the approach of old age either in my mental or physical powers".[4] On his seventieth birthday he was able to congratulate himself on being "still sound in wind and limb".[5] Six months later this country was engaged in a life-and-death struggle with Germany, inadequately prepared mentally, morally, and materially: a war in which the essence of the situation was that, though Britain and France together had superior man-power to Germany, and vastly greater resources to draw upon, we were caught at the time in a position of signal inferiority in both. And this after nine years of National Government, to Hitler's seven.

What is the explanation of this extraordinary passage in our history? It is, quite briefly, that the class of rich business men of whom Mr. Chamberlain was the ideal representative, and before him that other business man, Mr. Baldwin — the men who made the "National" Government and stood behind it all along — have been so paralysed by the contrary pulls of their class interest and the interest of the country, that they have ended, it is true unconsciously, in betraying the country they ruled. Ignominious and disastrous as their period of rule has been in time of peace, they were still less competent to lead the country in time of war: at the bottom of their failure was the fact that you cannot conduct a totalitarian war and preserve their system of private enterprise. But so entrenched were these men in power, throughout the Conservative Party, which had become their party, in the House of Commons, which had become their club, that it has taken disasters of the greatest magnitude to bring it to an end. Their fall represents a political revolution of the first importance, the significance of

[1] *Ibid.* p. 302. [2] P. 303. [3] P. 311. [4] P. 377. [5] P. 413.

The End of an Epoch

which has hardly been realised in the crisis of our fate in the war itself.

But it is surely of great significance for the future that the country to save itself has had to turn for leadership to a combination of aristocrats with the working-class movement. Chamberlainism has fallen, like Lucifer, never to rise again. The men of Munich and of these years that the locusts have eaten have been uprooted, scattered, some kicked upstairs, some out of office altogether, some sent abroad ignominiously to fetch home payment for the services they rendered. The men who have been right all along, Churchill, the Labour leaders, and the Conservative rebels, are in control of the war: it is they who occupy the key positions. And all this rests upon an anomalous position in the House of Commons, where the Conservative caucus and Mr. Chamberlain still have a majority. The House will have in time to be brought into proper relation both with the Government and the country. War-time Socialism has been brought into being overnight. There is a new spirit abroad in the nation. Yet all now awaits the arbitrament of war. An epoch of unparalleled ignominy in our history is ended, even though it has to be atoned for in blood.

VI
REFLECTIONS ON LORD BALDWIN
[1941]

"There is one other thing. The greatest crime to our own people is to be afraid to tell the truth."—Mr. Baldwin, Debate on Armaments, House of Commons, July 1934.

"My position as the leader of a great party was not altogether a comfortable one. . . . Supposing I had gone to the country and said that Germany was rearming and that we must rearm, does anybody think that this pacific democracy would have rallied to that cry at that moment? I cannot think of anything that would have made the loss of the election from my point of view more certain."—Mr. Baldwin, Debate on Defence of the Country, House of Commons, November 1936.

"The responsibility of Ministers for the public safety is absolute and requires no mandate."—Mr. Winston Churchill, House of Commons, November 1936.

"We never worry. Worry works internally and affects the nervous system."—Lord Baldwin, *The Falconer Lectures*, Toronto, April 1939.

FOR fifteen years, at a most critical juncture in our history, from the fall of Mr. Lloyd George in 1922 to Neville Chamberlain's assumption of the Premiership in 1937, Mr. Baldwin was the most powerful man in British politics. It has not been given to many of our modern statesmen to enjoy such a long, and almost unbroken, run of power. In fact, it would be out of the question for anybody who was not the leader of the Conservative Party, who had not behind him all the forces of which only the Conservative Party can dispose, the City of London, the banks, industry, the land — or, at any rate, the landlords, the farmers, and the bulk of their obedient tenants and labourers, most of the newspapers, the Church (that other organ of propaganda), the aristocracy, the middle classes, upper and lower, the infinite and incalculable influences

77

of respectability and social snobbery — in short, all the holders and supporters of power. Nobody who represented the other side, or — though he might be a man of genius — was outside that camp, had the slightest chance. But it happens that in Mr. Baldwin the English middle classes had a man after their own heart. No man of genius could have disputed his ascendancy with them. There were, in fact, two such men whom they could have called upon to lead the nation through those difficult and dangerous years: the great man whose courage and tenacity rode the storm of the last war and brought us through to safety; and the great man whose unbreakable spirit and resolution inspires us in the trials of this war which need never have come upon us if the country had but heeded his warnings. But the Conservative Party would have neither of them.

In 1922 it turned Lloyd George out of power, in the prime of his experience and prestige, the most famous figure in the English world, and in 1923 chose a quite unknown man, the "cabin-boy" instead of the "captain" — as Lord Birkenhead commented contemptuously but not inaptly — and held him in power and esteem throughout the whole of this period to 1937, in the course of which we lost all that we had won at a cost of four years of war in 1914-18 and at the sacrifice of a million British dead. Nor was that all: it landed us, after years of unchallenged power, in another war in far more dangerous circumstances, when all has to be fought over again, only twenty years after emerging from the first with complete and absolute victory. In incalculably worsened circumstances: last time, after eight years of Liberal government, the ablest we have had in modern times, when the war came we had France and Russia on our side, and Japan and Italy and the United States, to say nothing of smaller peoples. This time, after twenty years of Conservative rule, twenty of the most disgraceful years in our history, we are fighting for our lives, and virtually alone.

It is of the first importance to bring home whose responsibility it is to the right quarters, if we are to learn the lesson of the past for the benefit of the future. As Mr. Churchill has said: "We are not in a position to say 'The past is the past'; we cannot say 'The past is the past', without surrendering the future". Political leaders cannot abnegate responsibility for what happens to the country under their leadership; again, as Mr. Churchill has said: "The responsibility of Ministers for the public safety is absolute and requires no mandate".

In Mr. Baldwin the Conservative Party and the occult influences which through its agency rule the country, had a man after their own heart. Coming from the Western Midlands — that sluggish region of the country politically — not a far cry from Birmingham, *fons et origo malorum* of British politics in the past three-quarters of a century, here was no Welsh genius whose energy and vision might run counter to the interests of property, however indispensable they had found his services in 1914–19. Here was no Churchill with an historian's inconvenient understanding of the basic necessities of British policy in Europe and the world beyond. They wanted to forget; they wanted to make money; they wanted to return to the so comfortable world they had inhabited before 1914. So they chose "a person of the utmost insignificance", as Lord Curzon bitterly lamented. It is the nation that has had cause to lament it since. But from their point of view Mr. Baldwin had two great advantages: an industrialist himself, he could be depended upon to see things in terms that would suit the interests of "industry". He did: witness the General Strike and his role as saviour of the nation in 1926. (There would have been more point if he had acted as the saviour of the nation from the greater dangers of 1933 onwards; but that would not have suited the purposes of "industry" so well. The whole logic of the attitude of "industry" to Nazi Germany pointed to a deal with it; and it must not be forgotten that at the moment Hitler was invading Czechoslovakia, representatives of the F.B.I. were in Düsseldorf negotiating an understanding with their German confrères.) But a second, and hardly less important, consideration: Mr. Baldwin talked the language of the people; his was a plain, straightforward, bluff, honest, truthful personality such as the people could understand. These were democratic days. Mr. Baldwin's expressed sympathies were all in favour of democracy. Was he not himself a supporter of the extension of the franchise, especially to women? — in itself a great proof of democratic sympathies, while at the same time not at all a bad move for the interests of the Conservative Party, since women are much more conservative than men.

So they gave him their confidence as few party leaders have been given it — not clever men like Balfour or Disraeli before him, certainly not great men like Gladstone or Peel, Canning or Pitt. For Mr. Baldwin was an inferior man, and that appealed to the great mass of inferior people who are the electorate and love that sort of thing. They felt safe with him — when they did not feel safe with Mr. Lloyd George or Mr. Churchill. They were very happy and content to go on, year after year, with him, in spite of the warnings they were given. As a political candidate with a responsibility to the people in my own part of the country, the county of Cornwall, I warned them again and again of the fatal indolence and irresponsibility of this man who was Prime Minister of Great Britain. They preferred not to listen; they are paying for it now. But who am I to complain when not all the eloquence, the experience, the hereditary position, the great name of a Churchill had the slightest effect (and there were others: the best elements in the nation) on a lulled and fascinated and befuddled people? For years this man held office and power and respect and consideration. All that suffered was the long-range interests and the security of Great Britain and the Empire. It is impossible for him now to avoid responsibility. His career and personality, above all his technique of managing the electorate, and his appeal, are phenomena of the first importance to be studied for all who wish to understand the true character of British democracy and its workings in the past twenty years and who desire to avoid the mistakes, the fundamental fraud running through the whole system as it has been operated, for which we are paying so heavily now.

This is not the place to review his whole disastrous record,

Reflections on Lord Baldwin

but simply to concentrate on the evidence that his speeches afford us on those few cardinal points which are most relevant to the crucial question of leadership in a democracy. It is because of the failure there that we have to go through the ordeal that we are now going through. *It need never have happened.* It need never have happened had this man done his plain and simple duty by the people who were in his care — who trusted him as they trusted no one (more fools they!), whose trust he at every moment and in every speech solicited, flattered, conjured — and ended by betraying more completely than we have ever known in our history.

What did Mr. Baldwin really think about democracy, about the people who were such fools as to go on supporting him? What was his attitude on the crucial questions of responsibility and leadership?

Mr. Arthur Bryant in his official tribute to Mr. Baldwin tells us that " more than any public man of his generation, Mr. Baldwin has believed in democracy because he has believed in the common man. Like Cromwell and Lincoln, he learnt to know and love him " . . . etc. It is certainly the common man who is now suffering from the ardour of that belief in him, from the too close and stifling embrace of that love for him. Better the withering contempt of a Swift or a Strafford, a Voltaire or a Shaw for the common man — provided only there were some glimmer of a sense of responsibility for him, some care for his simplicity and credulity, some respect for his faith and loyalty, which treated him as he deserves to be treated for his dull, laborious days, warning him faithfully of his dangers, thinking ahead for him, telling him the truth about things, guiding him, protecting him, leading him; not using your wits, your cunning to lull him into a sense of false security, telling him that all is well so long as he trusts you, then letting him down, failing him at the moment of greatest danger. Better a thousand times an honest contempt for the people, and a respect for their defencelessness, than that kind of " democratic " humbug.

In all his speeches Mr. Baldwin always made great play with the idea of himself as a democrat, flattering the British

The End of an Epoch

people with constant appeals to their weakest side — their distrust of intelligence (so different from the great days of the sixteenth and eighteenth centuries when the foundations of the British Empire were laid, and so different, alas, in consequences then and today); constantly playing up to their silliest side, their love of moralising, their love of seeing themselves as neither clever nor logical, as other people are supposed to be, but honest and good; pandering to the gallery with his insistent depreciation of intelligence — as if this great country achieved its name and fame in the world by ignorance and stupidity; in the end posturing before the people, a caricature of himself, the old Harrovian who has talked pulp to the people until his mind has come to consist of the pulp in which he dealt so long and so successfully. The contrast it is to read yards and yards of these sickening speeches in Hansard with their debilitating egoism — their chief concern always being not the issue in hand, not the pros and cons of policy, but how he is appearing to the people, speeches intellectually invertebrate, without shape or style — and, on the other hand, the speeches of Mr. Churchill, always addressed to the matter in hand, devoting "fundamental brain-work" to the issues of policy upon which depend the well-being and safety of the nation, and at the same time admirably constructed, full of wit and humour, and inspired by a profound sense of responsibility and an historian's sense of the needs of policy. Yet the Parliaments of 1931 and 1935, and the people outside, listened to the former and would not hear the latter, until events made them hear, and with a vengeance.

For Mr. Baldwin told them, in every speech, always that *he* was the democrat. In the very speech in which he had to defend himself in the House of Commons for the trickery of the attempted Hoare-Laval deal with Mussolini only a few months after winning an overwhelming majority on the plea of pursuing a League of Nations policy, Mr. Baldwin said: "I shall always trust the instincts of our democratic people". This is the very speech in which he was forced to stand in the white sheet of a penitent for having misled them for months on this issue, as he had been misleading them for years on other

Reflections on Lord Baldwin

and even more dangerous issues. What is the clue to his behaviour? It is to be found in a speech in the House in 1934: " Mr. Keynes told me ten years ago that it is quite impossible to make the English people think, and that I was wasting my time, but I did make a great effort then ". Apparently he had desisted ever since. Now ten years takes us back to the election of 1924, when the Conservative Party, by its skilful and unscrupulous exploitation of the Red Letter scare, obtained a large majority for the next five years: which they used to return to the Gold Standard in the interests of the City, thereby throttling British exports, their remedy for this being to reduce wages and so force the General Strike — which they were able to represent as a danger to the nation and easily defeat, and to institute a deflationary policy under the aegis of Montagu Norman that set in train the world economic depression, while such time as they could spare from these activities they spent in neglecting the fundamental problems of European order.

But — and this is the point — the year before, Mr. Baldwin, who as an industrialist was, of course, a protectionist, had gone to the country for a mandate for tariffs and been defeated. After that, never again. He had learnt his lesson. Mr. Bryant lets the cat out of the bag. " He imagined he could show the creature [*i.e.* the electorate] its mistake by reasoning with it. . . . He did not know them as well as he supposed." Never again after that. The whole art of politics became for him the art of taking the people in, telling them what they wanted to hear, waiting to see which way they were moving, and then — the leader of the greatest party in the State, Prime Minister of Great Britain, responsible for the lives and security of forty-five million people — following safely, belatedly, cautiously after. " Safety First " was the motto under which he fought his elections and conducted his manœuvres: meaning safety for himself and for the Conservative Party. It must be admitted that he was very successful at that game; its ultimate consequences for the country are to be seen in our wrecked and ruined towns, in the City, in Westminster, in Birmingham, Coventry, Liverpool, Southampton, Portsmouth, Exeter, Plymouth — all of which supported him and his disastrous

The End of an Epoch

party through thick and thin,[1] and now must pay for it.

Henceforward, for the rest of his career, the country was to be governed by what can only be described as the consistent application of the confidence trick. He had found it worked so well; he was such a past-master of the art, with his appearance and technique. It reminds one of the story of the distinguished public servant who once came upon Mr. Baldwin alone in a room at some country house practising various angles at which to hold his pipe, before a mirror. His line at the time of the General Strike was really the confidence trick practised upon the country — only people had not come to recognise it then. "We must resist this unconstitutional threat to the nation, and once that is defeated you can trust me to see that justice is done to the miners." Justice never was done to the miners. The most ambitious and aesthetically complete example (Mr. Baldwin was an artist) of the confidence trick was the panic election of 1931. The old and worn-out hacks of the Labour Party were taken in, MacDonald, Snowden, the celebrated Jimmy Thomas: people are taken in chiefly because they want to be taken in, witness Hitler's technique with the propertied classes of Europe from 1933 on. Mr. Baldwin was very self-effacing; he was ready to surrender the first place in the National Government to the vainest of his dupes. It was cunning; he did not have to take responsibility openly — until the fact that MacDonald was gaga became too evident. Meanwhile, ensconced in the background, the unchallenged leader of the Conservative Party, the reins of power were safely in his hands and theirs. It was clever: it was not a Conservative Government that ran the country through all those years from 1931 on, until it ran the country on the rocks. Oh, dear me, no! — it was a National Government. Today the name of National Government stinks in the nostrils of all decent people, so that a Government of national safety, that really represents the nation, got together in the moment of greatest danger to save the country from the consequences of those disgraceful years from 1931, will not even take up the name from the gutter.

[1] I except Bristol from this stricture.

Reflections on Lord Baldwin

The confidence trick was played again and again, not only over great issues of policy, but over lesser matters, not only in the General Elections, 1935 as well as 1931, but over the Hoare-Laval swindle, most dangerously of all over the question of German rearmament with all its related and subordinate issues, the solemn pledges of air parity with Germany given by Mr. Baldwin to Parliament and the country, the holding-up of a promised Ministry of Defence for nine months and then the appointment of an Inskip as Minister — a typical piece of wool-pulling across the eyes of the public; the giving of mistaken figures as to German air strength in the House of Commons so as to throw doubt on Mr. Churchill's estimates, when the latter turned out to be more than justified in his warnings, the reiterated assurance that everything was all right, you could trust the Government.

> Baldwin's in his heaven:
> All's right with the world —

when everything was far from right, and Mr. Churchill was at length driven to describe the confidence trick in so many words. In the summer of 1936, when Mussolini had carried through the conquest of Abyssinia and was forming the Axis with Hitler to carry forward their joint aggression in Spain, Mr. Churchill said:

We are told, "Trust the National Government. Have confidence in the Prime Minister (Mr. Baldwin), with the Lord President of the Council (Mr. MacDonald) at his side. Do not worry. Do not get alarmed. A great deal is being done. No one could do more." And the influence of the Conservative Party machine is being used through a thousand channels to spread this soporific upon Parliament and the nation. But, I am bound to ask, has not confidence been shaken by various things that have happened, and are still happening?

That summer and autumn, while Hitler and Mussolini were now in a position to put into execution their far-seeing campaign of aggression against the British Empire, beginning with Spain, it became clear that all hope of air parity with Germany had vanished and that there was no likelihood

The End of an Epoch

of our catching them up in rearmament. Mr. Churchill initiated a debate on the state of the country's defences. He recalled Mr. Baldwin's specific pledge given as long ago as 1934, three years before:

> Any Government of this country — a National Government more than any — and this Government, will see to it that in air strength and air power this country shall no longer be in a position inferior to any country within striking distance of our shores.

Mr. Churchill underlined the fact that responsibility was never " more directly assumed in a more personal manner ". Yet the solemn pledge was never fulfilled in general outline, any more than the particular steps necessary to implement it were carried out. Sir Samuel Hoare, then First Lord of the Admiralty, was put up to answer: " I am authorised to say that the position is satisfactory. . . . We are always reviewing the position." Mr. Churchill commented:

> Everything, he assured us, is entirely fluid. I am sure that is true. Anyone can see what the position is. The Government simply cannot make up their minds, or they cannot get the Prime Minister to make up his mind. . . . So we go on preparing more months and years — precious, perhaps vital to the greatness of Britain — for the locusts to eat. They will say to me: " A Minister of Supply is not necessary, for all is going well ". I deny it. " The position is satisfactory." It is not true. " All is proceeding according to plan." We know what that means.

Mr. Churchill went on to give evidences that no attempt had been made to carry out the pledge or equip the Army properly: searchlight companies without searchlights, the Army lacking " almost every weapon which is required for the latest form of modern war. Where are the anti-tank guns, where are the short-distance wireless sets, where are the field anti-aircraft guns against low-flying armoured aeroplanes? " In the air we had already fallen more than a third behind Germany — three years after a solemn pledge to keep parity. He called upon the Prime Minister to account for his responsibility before the nation.

In reply Mr. Baldwin made one of the most ignominious

Reflections on Lord Baldwin

confessions that have ever fallen from the lips of a British Minister. He admitted that

from 1933 I and my friends were all very worried about what was happening in Europe. . . . But I would remind the House that not once, but on many occasions . . . when I have been speaking and advocating as far as I am able the democratic principle, I have stated that a democracy is always two years behind the dictator. I believe that to be true. It has been true in this case.

He went on to excuse himself on the score of the Fulham by-election in 1933, when the National Government lost one of the 550 seats it won by the panic of 1931.

My position as the leader of a great party was not altogether a comfortable one. . . . Supposing I had gone to the country and said that Germany was rearming and that we must rearm, does anybody think that this pacific democracy would have rallied to that cry at that moment? I cannot think of anything that would have made the loss of the election from my point of view more certain. . . . Frankly, I could conceive that we should at that time, by advocating certain courses, have been a great deal less successful.

So that was the explanation: the trickery of a party manager where there should have been the sense of responsibility of a Prime Minister of Great Britain, if not of a leader of the Conservative Party, for the safety of the nation. This barely concealed distrust of the people who were following him was the deepest slander upon them, as was seen from the way they rallied in the worst circumstances of defeat and the fall of France last year to the sincere and candid leadership of a Churchill.

In fact, nothing could be meaner than the attempt to shuffle off the responsibility from the shoulders where it properly lay — upon the country, upon the electorate, upon the League of Nations Union, upon Lord Cecil and Professor Gilbert Murray, upon the Labour Party — anybody rather than that party and its leader which had enjoyed the fullest power throughout that decade from 1931. They now say: " We were all to blame ". It is very generous of them:

The End of an Epoch

another variety of the confidence trick. Mr. Churchill has himself disposed of these excuses once and for all:

> What is this argument that it was necessary to wait for the General Election before Ministers could do their duty and place their country in a state of security? . . . It has been said that before the General Election there was no mandate to do it, but there is no mandate so imperative on Ministers of the Crown as that they should guard the safety of the country. Throughout the last Parliament there was never a moment when the Prime Minister could not have asked both Houses of Parliament to support him in any measures necessary to maintain the security of the country. Nothing relieves Ministers from that prime duty. It is the first object for which Governments are called into being.

Even earlier, Mr. Churchill had exposed any such excuse:

> I have been told that the reason for the Government not having acted before was that public opinion was not ripe for rearmament. I hope that we shall never accept such a reason as that. The Government have been in control of overwhelming majorities in both Houses of Parliament. There is no vote they could not have proposed for the national defence which would not have been accepted. . . . As for the people, nothing that has ever happened in this country could lead Ministers of the Crown to suppose that when a serious case of public danger is put to them they will not respond to any request.

On one of his tours of a blitzed town recently, greeted by the cheers of people who had been bombed out of house and home, who had lost everything, Mr. Churchill praised their spirit and said: " I can never deserve it ". But he had already deserved their trust, not only for his own magnificent courage and determination, but for the years in which he fought against that majority which nothing could shake, for the safety and security of the nation. During all these years Mr. Baldwin held power in fuller measure and with a larger majority than it has ever been given to a political leader in this country. On the fundamental test of the responsibility of leaders to the country whose destinies they guide, he is tried and found wanting. His whole career throws a searching light upon the crucial question for the future, that of leadership in a demo-

Reflections on Lord Baldwin

cracy. His often-protested belief in democracy is seen to be largely humbug; his trust in the people, in the literal sense hypocrisy, even if he himself was not aware that he was playing a part; his career of "love and service" to consist mainly of keeping the country safe for the Conservative Party. On rule by the confidence trick, that most penetrating of political observers, Low, is at one with Mr. Churchill.[1] It is true that rule by confidence trick prevails in totalitarian States where there is no criticism. But it is still more dangerous for a democracy, which depends for its very existence upon a genuine and unfalsified political life, and where this kind of rule does not even achieve power.

Mr. Baldwin, now elevated to the peerage as Lord Baldwin, lives still in possession of his honours, if not of honour, and presumably of a pension for his service. But what can this man think in the still watches of the night, when he contemplates the ordeal his country is going through as the result of the years, the locust years, in which he held power?

[1] *v.* below, p. 110.

VII

THE WORLD AND U.S. POLICY

[Review of Sumner Welles, *The Time for Decision*:
Sunday Times, Oct. 29, 1944]

THIS is the most authoritative and important book on American policy towards the outer world to come from the United States; it is also a most valuable and revealing contribution to the history of our time. Mr. Sumner Welles has been at the centre of affairs in the State Department at Washington for the past ten years, and before that a *diplomate de carrière* whose work took him to Tokyo, to South and Central America, and acquainted him pretty thoroughly with Europe and the sources of our discontents. His book is a convincing testimony to the value of professionalism in these matters; he really understands the causes of our troubles. A man of ability and penetration, imbued with strong convictions and principles, he is that rare thing, an exponent of a liberal-minded internationalism whose feet are firmly on the ground and who never for a moment loses sight of what the world is really like. Add to that a refreshing, an undiplomatic candour and you see how revealing and instructive a book this is.

It falls into three parts. The first may be described as " What Might Have Been ". It portrays the world scene from the end of the last war to America's entry into this. It lays bare the two foci of aggression in Germany and Japan, the germ-centres of an irreconcilable mentality with which, we have learned, it is impossible for the rest of the world to live in peace. Theirs have been deliberate sins of commission, crimes against the human race. But Mr. Sumner Welles is no less severe on the sins of omission for which the democracies are responsible. His book is written for the American public, and he lays blame fairly and squarely on democratic opinion for refusing to back up the positive policy of its Government, if necessary, with force.

The World and U.S. Policy

That is the whole moral of the book. Mr. Sumner Welles believes that if the democracies had been willing to back up their peace system with force, after the experience of the last war, all the disaster of this might have been avoided. And I think he is right. The essence of the situation has never been better expressed, to my mind, than by the American admiral who said that " the power to wage war must be in the hands of those who do not believe in war ".

This is the central problem of policy for democratic government; and the very fact that the democracies are peaceably inclined led to their undoing. Mr. Welles says outright that nothing was more truly harmful in its effect, in its encouragement of wishful thinking, than the Kellogg Pact. When one considers how President Roosevelt was hampered at every turn in his attempts to preserve peace, by the isolationism of the American public, by the Neutrality legislation imposed by Congress, it is wonderful what he managed to achieve: he emerges not only as a great statesman, but a true friend to the world. The only solution for the democracies is to educate them to their responsibilities, and what an education by experience they have had!

Naturally the criminal forces were enabled to get ahead meanwhile. And not Nazis only. Mr. Welles regards the whole German campaign since 1918, the systematic evasion of disarmament and reparations, the building-up of force for a war of revenge, the working-out of a system of " indirect complicity " by which all kinds of people, often unconsciously, subserved German purposes in their own countries until these were undermined from within — he regards all this as the long-term plan of the great General Staff, the real brain-centre of the German people. One hears a lot about the " good Germans ". But I, for one, did not know that Stresemann, the good German of Locarno, offered France a German alliance behind our backs, really directed against this country.

Mr. Welles, with all the authority of his knowledge and experience, regards German unity as " a continuing threat to the peace of the entire world ", and he is in favour of the partition of Germany. With great respect to his judgment, I do not think this would work: it would give the Germans a united objective to work against: they would undermine it from within, and to allow them to succeed would be disastrous to any peace settlement. We should probably obtain as much from insisting that Germany becomes once more a Federal State, instead of a State over-centralised for aggression.

But Mr. Welles is utterly right in pointing out how much less propitious for democracy Germany will be after this war than in 1918, with a whole generation of young brutes growing up since 1933 to maturity. " These millions of Germans will be at the prime of life during the next two decades. They will be a controlling force within Germany. Theirs will be a force of fanaticism and revenge."

There lies a problem that we cannot escape — though no one seems to be thinking about it.

Mr. Welles gives us a fascinating account of his mission to Europe early in 1940, before the storm broke on us. There is all the macabre atmosphere of Berlin, the prison-yard of Hitler's Chancery, the brute-faces on guard, the flowers within, the secret dictaphones installed in the walls of the Foreign Office. There is Ribbentrop " saturated with hate for England ". Paris was already worm-eaten with complicity, direct and indirect. The contrast in the spirit he found in London wins a well-deserved tribute: " In England there was no evidence that I could see that German propaganda, which had so fatally undermined French morale, had made any headway ".

The last part of Mr. Welles's book is a plea for an international organisation continuing the alliance of the United Nations and resting upon regional associations of powers.

Years of writing memoranda in the State Department are hardly conducive to a gay *insouciance* of style; but the importance and the convincing character of what Mr. Welles has to say give his book an absorbing interest.

VIII
DEMOCRACY AND DEMOCRATIC LEADERSHIP
[1941]

"The world at large wants a clear lead, and it is from the artists, writers, and teachers that plain directives must come, if they are to come at all."—H. G. WELLS.

(1)

THE trouble with so much of our talk about democracy is that with the talk the concept has become progressively emptied of meaning. We are supposed — and as I think rightly, if we understand all that it implies — to be fighting this war for the cause of democracy. Yet I doubt if there are many who are at all clear about what is meant by it, or are in fact roused to enthusiasm by the name under which they fight. So many of those who think must have been depressed by the hesitations, the incompetences, disillusionments, fraudulences of the past decade — the most disgusting and disquieting in our latter history, that has led us not surprisingly to where we are today — that they may think of the standard of democracy as rather moth-eaten. Yet they would not be wholly right, in spite of so many evidences on the surface of our political life. We must, in short, do some new and plain thinking on this subject, disregarding the clichés and soft illusions with which the " supporters " of democracy have coddled themselves and us into so many defeats that need never have been. If we are to win, we must at least have the intellectual courage to face unpleasant truths about our own cause.

It is only comparatively recently that the meaning of democracy has become so vague — as vague and diffuse as " socialism ", or " liberty ". The nineteenth century, to take

The End of an Epoch

such admirable writers as Cornwall Lewis or Bagehot, attached a perfectly clear meaning to the term: they meant, as Aristotle meant, a form of government determined by the predominance of the propertyless Many as against the Few — and as such they feared and deplored it. (They need not have feared it so much: little did they realise that as the franchise was widened to include the bulk of a politically immature and uneducated people, the influence of property upon the political life of the nation did not necessarily decrease, nor the influence of money upon the results of elections.) As the franchise was extended, the meaning of democracy became more diffuse with it, and people were ready to accept Abraham Lincoln's befoozling " government *of* the people *by* the people *for* the people " as not only the last word in political wisdom, but as telling them something about the nature of democracy. But Lincoln was not a political thinker but a practical politician; he knew how important it is to tell the people what they like to hear.

Government can never in our large nation States be *by* the people in any effective sense; it may be — and since we are democrats, should be — by their true representatives, those who represent their real interest, the general interest of the community. Those are not necessarily the majority. I can conceive of cases — have we not the Hitlerite plebiscites before us as an example? — of a Government getting a 95 per cent vote of the people and yet not representing that people's true interest; conversely, it may be a small minority who really express the general interest. It so happens that in the last decade the " National " Government had an immense majority of the country with it at the General Elections of 1931 and 1935. But the whole course of their actions showed that they did not really represent the interests of the country. (The very fact that they knew this themselves operated in a subtle way to paralyse their action.) Events caught them out in the end; if the Conservative Party, which was the reality of power behind the façade of " National " Government, with the propertied classes behind it, had but done its simple duty by the interests of the country and people it purported to represent we should not now be fighting for our very survival as a nation.

Conversely the forces of the Left, centring upon the Labour Movement — though they made every conceivable mistake, quarrelled incessantly among themselves, refused to unite or even work together in face of the gravest danger, put their second-rate figures up as leaders for the country to follow, made no effective use of the immense resources of ability and devotion among the younger generation, had no sense of effective popular appeal, no sense of propaganda, no sense of power, and were like what Henry Nevinson said of the (Tsarist) Russian Army, that " had no sense of time or space : it was always late and never knew where it was " — yet, in spite of this deplorable record, the simple fact is that the Left represented the true interests of the people of this country, who voted them down practically every time they presented themselves for a mandate to govern. Events proved that to be true : in politics the proof of the pudding is in the eating. It was the series of disasters and betrayals and defeats associated, as it will be for ever in our history, with the name of Chamberlainism, that forced the Churchill–Labour Government into power; not a majority in Parliament, which was indeed against it. Not for some time did a majority come round to it in that *Chambre introuvable*, that deplorable assembly elected by the fraud of 1935 and never yet renewed. An anomalous situation, to be sure, to find in the name of democracy !

And yet it is hardly exceptional. During the twenty-three years that elapsed since the last war this country has been continuously governed by the Conservative Party — the political instrument of the propertied classes — save for two brief intervals in 1923 and 1929–31. The reason for the deterioration of our political standards, for the systematic befogging of the public mind on issues of the gravest importance, for the sickening insincerity and living hypocrisy which was the very nature of National Government (Baldwin-MacDonald-Chamberlain-Simon-Hoare-Runciman : how a common character runs through them all), was simply that their position was a false one from the very beginning : their whole pretence was that they represented the bulk of the nation — and the bulk of the nation was fool enough to believe them, until events taught them in the blood of their sons.

(II)

What light do these things throw upon the nature and working of democracy in this country? Can we be said to be a democracy at all? What are we to do to reform matters, and not merely to put them right, but to bring them into relation with the claims we make for ourselves, the cause for which we are now fighting? How to explain? How to relate the facts to democratic theory?

It is not difficult. Rousseau as usual goes to the heart of the matter with a clear, candid generalisation. He says, speaking of the general will in a society that a Government should represent: " What generalises the Will is not so much the number of the votes as the general interest that unites them ". The central problem of democratic politics becomes then how to bring the voting power into relation with the general interest of the community.

The plain and practical answer to that, and indeed the obvious deduction to be drawn from all that has happened in our internal politics in the past twenty years, is that we must at least *clear the channels*, so as to give democracy a chance of working at all. It is an absolute *sine qua non* of democracy existing. As things have been, the real interest of the people has been thwarted, frustrated, deflected, vamped, seduced, their votes solicited for the benefit of others, bought indirectly if not directly, influenced when they were not commanded by the overt economic power of employer or landlord: in short the electorate, or a sufficient amount of it to turn the scale, influenced in a hundred different ways which it would take a whole volume to describe.[1] Unfortunately intellectuals, under the influence of the rationalist fallacy and with no experience

---

[1] I suggest this as a more profitable subject for political research than most undertaken, for example studies of Public Opinion. As if Public Opinion were an independent entity!

of practical politics, are apt not to appreciate the decisive importance of these methods and influences. But the fact is, *they must be stopped if we are to achieve an effective democracy, the channels cleared for the real interest and the will of the people to flow rightly and naturally.*

Most of our political system, in some cases the institutions themselves, in others the way they operate, is arranged in favour of the existing order. The very complexity of it for one thing is a factor holding up progress. It is all very natural and historical: these institutions come down in a long tradition from an earlier social order. It is equally natural and in the logic of history that they should be brought into relation with changed circumstances. Our political institutions, in short, will need thorough overhauling when the war is over; they will need to be greatly simplified and made more directly responsive to the interests of the great bulk of the people. Take a few examples. In case the Left should by some mischance win a General Election, there is the House of Lords to prevent it carrying its policy into action. It needs recasting: those who have not studied the subject would hardly believe its record of effective obstruction and mauling of progressive measures.[1] (We owe the poisoning of our relations with Ireland largely to the action of the Lords in holding up Home Rule and remedial land legislation in the nineteenth century.) Take the anomalous relation of the State to the Anglican Church: it would be an absurdity, as it is certainly an anachronism, if it did not play some part in helping the existing order to stick together. There is no intellectual defence for it:[2] the proper place for churches in the modern community is as free and voluntary associations for those who want that sort of thing, and those who want it paying for it.

Take again the structure of local government. The county councils have had enormously increased powers conferred upon them. Yet, with two or three exceptions — London, Durham,

---

[1] I have studied this record in *The Question of the House of Lords*. Hogarth Press.

[2] As the candid ex-Bishop of Durham recognises; cf. H. Hensley Henson, *Disestablishment*.

Glamorgan — they are entirely in the hands of the propertied classes, and it is virtually impossible for the Left to have any voice on them, let alone effective representation. Working men can afford neither the time nor the money to sit on them; there would not be the men even if they could. It would be far more representative and truly democratic if — once the mandate is given by the people at a General Election for a Left Government at the centre — that Government were then to appoint its representative in each county to administer it with the aid of appointed advisory committees. The ending of the county councils would raise not a murmur: they have no real existence in the will of the people. Smaller local bodies like town and parish councils have a more real existence and are more truly representative in character. They might be retained — though everybody knows there is a certain deterioration in public interest, and perhaps even in standards, in these local bodies.[1] With the far greater complexity of modern government, their affairs are increasingly run by their officials. The important thing is to organise on a national basis the municipal civil service, give it national status and standards, bring it along with other black-coat workers, for example the teachers, into their proper place alongside the workers organised under the T.U.C. (It would be so good both for them and for the T.U.C.)

Or there is the electoral system. Only those who have had experience of it as candidates or party workers can have any idea of how it is weighted in favour of the existing order. The Conservative Party caucus has got a fund of several millions at its disposal collected by selling honours and in other ways. The influence of that is enormous: they can buy all the posting stations, put forth millions of posters and leaflets, organise mass canvassing. In the counties they possess all the cars on polling day; in the towns and villages practically all the best party premises and clubs. There are all the innumerable influences of social snobbery at their disposal, in addition to the large voting bags of the domestic-servant class and the agricultural interest. It is not for nothing that the

[1] Cf. an interesting article in *Fabian Quarterly*, 1941.

earth is theirs and the fullness thereof. Indeed, when you consider the range and importance of these forces under their control, it is a wonder that the Labour Movement has any electoral support at all. (A tribute to the goodness of human nature, perhaps.) One may reflect that after the war is over — what with the sale of securities abroad, the enormous diminution in the returns to the rentier on investments, increased taxation, etc. — a good deal of this will be changed anyhow: there will not be so much money about for them to do what they like with the electoral system.

Again, there is the press to be considered, not merely as an industry, but as a social institution. It is intolerable that after the war a handful of millionaires should continue to hold the irresponsible power and exert the nefarious influence they have done, cheering Poincaré and the Comité des Forges on in their Ruhr enterprise against a Republican Germany, supporting Japanese aggression in Manchuria, pushing appeasement with Hitler and a Nazi Germany, backing Mussolini as the Conservative press did ever since he came into power right up to Abyssinia and even after, supporting Franco against the friends and the interests of this country. That cannot be tolerated if this country is to be an effective democracy: democracy does not mean *laissez-faire* for press magnates any more than it means *laissez-faire* in industry and economic affairs. Public control of the press by a Government which really represented the interest of the community — it might be organised as a corporation with a certain amount of independence like the B.B.C. — would be far more democratic than the present antiquated feudalism of the press, with the press lords as the " over-mighty subjects " of today.

All these political and social institutions do but reflect the fundamental character of the economic system. That will be brought under public control; it is very largely so, under various forms, now as in the last war. But at the conclusion of this, it will not be possible to scrap it as they did after the last war. The whole thing has gone too far and become too necessary for industry to be able to run without it. Effective democracy requires socialism in economic matters to complete and implement it. That is not to say any of the idealistic moonshine of the Guild Socialism or Workers' Control variety: the public control of industry requires varied forms of organisation to suit the circumstances of the particular case, leaving plenty of room where appropriate for private enterprise in the interstices of the system.

(III)

It can be seen what a work of clearing the ground and of rebuilding there is before us. To tackle it at all adequately, we shall need fresh thinking, a new mentality untrammelled by the clichés that have rendered social democracy so ineffective in the past. To make democracy effective there are three essentials: Leadership, Propaganda, Education.

I notice that Priestley is a little frightened of the word "leadership". He need not be: it is not leadership that we have had from our ruling classes in the past twenty years, but systematised fraudulence. He must not be: for people have not caught up with the fact that there has been a revolutionary change in politics since the nineteenth century: where then there was a small electorate of a million and a half, of a fairly high standard of political intelligence, there is now an electorate of twenty millions, whose standard of intelligence and outlook is pretty accurately gauged by the popular press and the films. In their own interest they must have leadership. To abdicate the task is fatal; it only means that the job will be undertaken by the gangsters or their gentlemanly collaborators. That was the fatal mistake of social democracy, not to give the masses leadership, to have no sense of appealing to them, no imagination, no sense of power. The Fascists have had all this — hence their success — and used it for criminal purposes; we must be just as effective and skilful, for the right purposes: we know what they are: we are agreed about that. Nor do I mean by leadership the *Führerprinzip*: I mean the group leadership which organises all the effective nuclei throughout the country, the ones and twos and half-dozens — hardly more than that — in each town and village, every shop, factory, school, and university, those who possess the indispensable gift of political initiative, infinitely rarer than intelligence — though I do not wish to depreciate intelligence. These people — it may be a china-clay worker in Cornwall, a bus conductor in Oxford, an R.A.F. mechanic — will often astonish you by the soundness and grasp of their political understanding, where the judgment of intellectuals is hopelessly at sea. These are the people who indicate themselves: it is they whom we must bring together and appeal to: they must understand each other, for there the real leadership in a democratic community is to be found.

Our message must be directed above all to them, helping to clarify their objectives, bringing them together, particularly since they belong to the generation that will emerge from the war; above all keeping in touch with them. This is merely part of the function of Propaganda. Social Democracy again had no conception of the cardinal function of propaganda in the modern State (as in all States); it idiotically left the devil all the best tunes. Is it any wonder that it was defeated? Anyone might say that it was just asking to be defeated, and serve it right, if it were not that the consequences were so tragic. But it is clear that a modern State can no more dispense with a Propaganda Ministry than the medieval State could with the Church — which was indeed its Propaganda Ministry; nor is it likely that we shall be able to dispense with our propaganda organs after the war any more than we shall with the public control of industry. The real point is to see that the right people get hold of them and use them for the right ends.

No less important than Propaganda — though not more so as liberals and rationalists are apt to think — is Education. It is obvious that our educational system is at a standstill. And no wonder, for the direction in which progress is to be made entirely depends on what happens to the social system. As organised at present the educational system exactly corresponds to the social order: the elementary schools for the working class, the secondary schools for the lower middle class, the public schools for the upper middle class, Eton for the aristocracy and the rich. In the existing stalemate the churches are trying to rush in and steal a march on the people while their attention is directed to the war, to put across a body of worn-out doctrine that no longer represents what the thinking part of the nation thinks, and impose it on the schools and teachers. It is not my business to deal with education here; but the right principle is that the educational system should be brought into conformity with the changed and more egalitarian society that will emerge from the war.

Wells asks for directives: here are some. They need not be agreed to in detail, but I should expect them in their general aims to represent a considerable body of agreement.

It is now possible to answer our questions. If something like these things are achieved, then we may call ourselves effectively a democracy. Until they are achieved there sounds something equivocal in the name, and one cannot expect to be roused to enthusiasm by it. All the same, those objectives are implied by the cause we are fighting for; if it were to be defeated, we need not bother ourselves about such objectives. We are a democracy already, in the sense that our political institutions give us a chance of changing the political and social order if we make up our minds to change it. That is a cardinal difference already from totalitarian systems; no amount of voting can change them, only blood and force, defeat in war. We already exemplify those ultimate values that are the glory of democracy as against the brutality of totalitarianism: freedom of thought and expression, the seed-bed of art and science and culture, the belief in the value of the human personality, which of all the great contributions of Christianity to the world may well prove the greatest and the most enduring. It may be that if we hold on and remain faithful, we may succeed in this country in giving a new lead to Europe: a more human and more hopeful conception of a society at once democratic and socialist than the morose example provided by Moscow that has so divided and misled us since 1917.

IX

THE PROSPECTS OF THE LABOUR PARTY

[1937]

MANY are the analyses and examinations, sometimes of a would-be post-mortem character, that have been devoted to the Labour Party since 1931 : that year of disaster, the real dividing-line in contemporary Europe, when the hopes of the post-war period came to a resounding end and everything from that moment went wrong. The crimes of 1933, 1934, 1935, 1936, 1937 — when will it end? and *where* is it all going to end? — only followed the defeat of our own working-class movement in 1931, the complete victory of our governing class, and the betrayal of all hopes of a new order of public control within the nation and without, of internal economic security and collective security in Europe and beyond. That is how the historian of the future will see these tragic years since 1931.

But the present is the very moment for a survey that shall be detached as well as addressed to the present situation, that attempts to see the thing objectively and to take a long view. It needs to be independent and candid, avoiding the professional optimism of the party member and the clichés that pass for thinking among so many engaged in active politics. For the Labour Party is at an important turning-point in its history. It very much depends on the line that it takes and the possible developments in the next two years, whether it emerges with a mandate to govern this country at the next election. If it does not, then we may roll up the map of Europe so far as the Labour Party is concerned. It may continue its depressing minority existence into the dangers of the unknown future ; but it would have no power to influence the course of events on the European scene. Whereas a Labour

The End of an Epoch

Government, with power, in Great Britain — and it is not beyond the bounds of possibility — would be a major factor in world politics, and rightly handled, the chief means of effecting a new disposition of forces in the world.

At the moment there is a new spirit abroad in the Labour Movement after the success of the Bournemouth Conference, which meant more than merely that a successful Conference was held. It means that the party is now moving along the right lines, that there is an increasing integration of its forces after the weary time of mutual suspicion and distrust between Transport House and the Left Wing, dissipating the energies of the party in sectional struggle.

It is *wonderful* how good the Labour Party is at giving points away to its opponents — never was there such generosity in a political party; it insists upon washing *all* its dirty linen in public. You do not find such idiotic generosity in the Conservative Party: they never give points away. Of course, all this is a function of the political immaturity and incompetence of a working-class movement; but, in fact, the intellectuals are worse even than the working-class elements.

However, much was done to eliminate the sources of these discontents at Bournemouth: the hard work that the Executive had put in during the last year, the fruitful labours of the Commission to investigate the distressed areas, the evidence that the Executive was working well together; the willingness of the Trade Unions to meet the wishes of the constituency parties as to representation on the Executive, the increased representation the latter has got, with the consequent election of some representatives of the Left Wing; the dissolution of the Socialist League and the cessation of the distracting " Unity Campaign " with the I.L.P. and the Communists, evidence of the wish of the Left Wing to pull with the party, at last, and not against it. All these things are the beginnings of a new impulse in the party, that, if taken at the tide, may lead to victory and power. (The Labour Party, like all working-class movements, has not the sense of power — is not power-minded enough. That is a fundamental source of their ineffectiveness; they leave all that to the upper and middle classes, who have

The Prospects of the Labour Party (1937)

nothing to learn about the importance of power in politics. *They* know that that is what politics is about.)

Yet these new signs, satisfactory as they are, are only beginnings. Everything depends on how they work out in the next two years, whether the Labour Party is carried forward to power or no. And no one can doubt that as things are, as yet, the position is anything but satisfactory. Here we are after six years of National Government, after six years of disastrous misconduct of our foreign policy, with the evidences of ruin all about us, in the Far East, in the Near East, in the Western Mediterranean, on the home frontier (which Mr. Baldwin said was upon the Rhine, but we have not even been able to keep Belgium firm as a buffer against Nazi Germany — or Portugal to the " old alliance " if it comes to that; and are now reduced to offering colonial and other concessions to Germany's blackmail which we should not have dreamed of making before 1914 [1]); here we have all these evidences of the ruin of our policy since 1931 — and the hold of the National Government is not even in question; nor is there any immediate prospect of its being shaken.

There are reasons for that, apart from the incompetence of the Labour Party. There is the very natural and widespread fear of war in the country; and human beings in the mass are such idiots that the nearer the victory of their upper classes brings them to war, the more they turn to their upper classes to save them. That is instinctive; that is only human. But over and above basic psychological tendencies of this sort, the Labour Party has little to be proud of in the opportunities it has missed and the showing it has made in this period. Too often it has been a case of " snatching defeat out of the jaws of victory ", as *The Times* has unkindly put it. There are reasons again for this, chiefly the impossible situation that the party was manœuvred into over armaments, and the uncertainty of its leadership right up till the past year. But it is not enough to offer excuses. It is true that the electoral disaster of 1931 returned only a miserable remnant to the Parliament of 1931–5, and left the leadership of the party to a Lansbury.

[1] Cf. the early chapters of Grey's *Twenty-Five Years*.

The End of an Epoch

The point about the Labour Movement is that its inherent sentimentality is such that millions of its members would prefer that kind of leader. But parties do not win power by following the impulses of their ridiculous hearts.

To be candid: though there are a number of Labour Members in Parliament who are very well equipped to deal with particular subjects, coal mines, questions of local government, unemployment assistance, health, and so on, it is well known that there are comparatively few whom the party can rely on to debate matters of high policy, particularly international policy — which is of such overwhelming importance in these years. That may to some extent account for the impression that undoubtedly exists, an objection that one constantly encounters in many circles, often well-inclined, that the Labour Party has not the men, for one thing, with whom to govern. I think that this impression is perhaps exaggerated, though there is some point in it to which the party should attend. For though the Tories have at least as high a proportion of the stupid, if not of the downright incompetent, among their ranks, it does not so much matter to them, for they have quite enough of a staff to govern the country, and they have the sublime confidence in themselves necessary for governing.

If the Labour Party is ever to govern, it must get rid of this miserable inferiority complex of an oppressed class, which hampers it at every turn; in order to govern, it is necessary to develop the mentality of a governing class, not of the governed. The trouble, of course, is that the Labour Party springs out of the governed, inferiority complex and all. That is why it is so necessary to it to attract recruits from other classes. Nor does that much matter to the Tories, when they are, anyhow, administering the country on conservative lines, which means doing nothing very much. But if the Labour Party, which is professing to change the economic system and institute large social changes at a time of general crisis like the present, plainly suffers from a shortage of men of ability to do it with, is it any wonder that people remain unconvinced and sceptical? The party might well learn to administer its own affairs more ably than it has done in the past five years, before offering to

The Prospects of the Labour Party (1937)

administer the country. It is obvious that this question of *personnel*, of the right disposition of its resources, of the leadership of the party — bound up as it is with the confidence of the country, and contingent as that confidence is upon the right solution of the question — is one of first-class importance to the party. Yet it is hardly ever mentioned in public; it is doubtful whether its crucial importance is at all appreciated by the party's organisation, or whether it has even penetrated the heads of party members.

Again there are explanations and excuses. So many people of ability are stuck in their occupations or in their professions, or are married and so cannot risk politics for financial reasons. These economic disabilities weigh heavily upon a party of the poor. Conservatives have no such difficulties; the greatest nincompoops can buy their way into a seat, or bribe an easily-befooled electorate — for that is what most Conservative candidatures come to, deluding fools with all the barrage of inducements that superior economic means can provide. On the other hand, the Labour Party does not make the best disposition of what resources it has got — all the more stupid of it when those resources are so few. Look at the candidates it puts into its safe seats! Not that that is the fault of its organisation, nor entirely due to the dominance of the Trade Unions. It is largely due to the incurably democratic character of the party; almost any local Labour Party would prefer a sentimental fool for its candidate to a man of real ability. (And yet Mr. Attlee has " great faith in the wisdom of the rank and file ".[1] No great political leader has ever had much " faith in the wisdom of the rank and file ". A leader may have affection for and even trust in the rank and file; but what the masses want from a leader is — to be led.)

It is not that there are not the men available, though they may be young. (That in itself should be the greatest advantage. Look how the Nazis and the Italian Fascists have made their fortunes out of the appeal to youth — for the worst of all criminal causes, death and destruction.) It is extraordinary

---
[1] *The Labour Party in Perspective*, p. 136.

how the very ablest of the young men at the universities — after all, the seeding-ground for the political leaders and administrators in the past and right up to the present — are drawn into support of the Labour Movement. But the Labour Party has, so far, made extremely inadequate use of these resources. One must not criticise unfairly; for the very ablest of the Labour leaders have a good record in this respect. Among Trade Unionists, Mr. Bevin, whose own abilities are of the highest order, has always done his duty by the universities; and Mr. Morrison, Mr. Attlee, and Dr. Dalton are beginning to pay attention to the possibilities here. Yet the party *must* bring out these people, it must make a proper disposition of its resources, if it is to win the country.

This point must not be held over-severely against the party; for it looks as if the situation is improving in this respect, and the question should be solved with time. The case of the great Liberal Government of 1906 is very much to the point. For its ability and strength were entirely unexpected. It was a Government of young men without experience and with hardly any name in the country. The Liberal Party had for some years been split from top to bottom; the Conservatives had been in power for twenty years and were very disrespectful of the new Cabinet and its power to govern. Yet this Government proved one of the ablest we have ever known. There is hope for the Labour Party yet! What is wanting is another 1906. Yet that depends on the party making full use of the opportunities that are presented to it.

For success in politics depends largely on circumstances. In politics you can never create an opportunity that does not exist, however good you may be. The Left Wing never understands this; and this completely vitiates its political judgment. Political success is a function of the right circumstances and the capacity to exploit them to the full. The chances of the Labour Party depend chiefly upon two external factors: the foreign situation and the possibility of a slump. If there is a European war in which we are engaged, there will clearly have to be a really National Government, instead of the present sham-National Government, to fight it.

The Prospects of the Labour Party (1937)

If the foreign policy that has been pursued since 1931 results in a first-class *débâcle* for this country — which is not impossible: all the possibilities for such a *débâcle* are there — then vast prospects for new combinations are opened up. The actual defeat of this country in a foreign war, as the result of the equivocal policy pursued by the governing class since 1931, would mean our destruction. The defeat or the subsidence of either Hitler or Mussolini would mean the revival of the Left all over Europe. That is what our governing class, which is the longest-sighted in Europe, well sees and is determined to prevent. That is why they are prepared to make concessions even at the expense of British national interests. The point is how far will they go; if they sacrifice the interests of the country too far in the interests of the existing order in Europe, there may come about a reaction. There are very interesting possibilities of Conservative unity becoming somewhat damaged over colonial concessions: not all Conservatives want to see the German colonies handed back for the *beaux yeux* of Hitler. The Labour Party should be on the alert for all these possibilities in the foreign situation: they may in the upshot give the initiative to the Left once more, which has been wanting since 1931.

But why has the Labour Party not been able to capitalise the failures, the insincerity, the disaster of the Government's foreign policy? It is a most paradoxical situation: a series of retreats carried out all over the world to the applause of the quondam imperialists; Conservative opinion and Conservative papers applauding British defeats, or what comes to the same thing, the victories of our rivals.[1] The only thing that explains the policy of the National Government in all this is that our upper classes are following their class interests and sacrificing the interests of the country. Why has the Labour Party not taken that line and brought that home to the country?

For a variety of reasons: because of the time it takes to get a stupidly democratic movement to change course even

---

[1] Sir Norman Angell deserves great credit for being the first to diagnose this strange state of affairs, in his article "The New John Bull" (*Political Quarterly*, July 1936); see also his *The Defence of the Empire*.

The End of an Epoch

in the direst necessity: in the Labour Party you have to *persuade* many people incapable of reason. Persuasion is a rational process applicable only to intelligent beings. The Conservative Party never bothers about persuasion; it hardly even tells its followers what it intends to do; quite rightly, it is the business of followers to follow, not to raise objections. The whole psychology of the thing has been brilliantly grasped by Low's cartoon, in which Mr. Baldwin has brought the mule — or is it an ass? — of the British electorate to the edge of the precipice, and then utters the consoling words: " Be reasonable. If I hadn't told you I wouldn't bring you here, you wouldn't have come." In that sentence is expressed the whole national fraud of the election of 1935. But why has the Labour Party been so ineffective in showing it up? In addition to the slowness of a democratic movement in the uptake, the regrettable necessity to persuade people who ought not to have to be persuaded at all, there has been the doctrinairism of the Left Wing and the pacifists, which has played straight into the hands of the Government. Time and again in the last few years there has been a line-up of George Lansbury, Maxton, Lords Ponsonby, Noel-Buxton, and Allen with Lord Lothian, the Astors, Sir Arnold Wilson, and *The Times*. Lansbury's pacifism must have cost the party a dozen seats at the last election. The Conservatives estimate Sir Stafford Cripps's idiosyncrasies as worth thirty seats to them; perhaps this is an over-estimation.

Over and above this the Labour Party itself has taken the wrong line about Rearmament. They hesitated, and hovered and havered, until it was too late. People only understand a straight and simple line. I am told that in the by-elections people have been saying of the Labour Party that it was " against the country being defended ", that it " didn't want us to protect ourselves ", etc. I can well believe it; that is exactly what people would think, however unjustified we know it to be. And however foolish they are, there is some justification for their fear: for supposing we do not provide for the country's security, on the ground that we cannot trust a National Government, responsible for the disaster of our

The Prospects of the Labour Party (1937)

policy, with armaments; it sounds reasonable enough — but, in fact, if a Labour Government came in to take the right line about collective security, it could not suddenly call up air fleets and battleships out of the ground overnight to deal with Hitler and Mussolini. It is that consideration that makes the Left Wing standpoint on armaments such dangerous lunacy.

Of course, we cannot resist or obstruct Rearmament. I have known that all along, and said so ever since 1933. The Labour Party was right to attack the National Government for not trying to get disarmament in Europe while there was still time; they never meant business by the Disarmament Conference: the greater is the tragedy now.[1] But from the moment that Hitler was firmly in the saddle — and the English upper classes have done their level best to keep him there — it was all over with disarmament. I knew that; Mr. Baldwin knew that; but he kept quiet so as to be sure of the election for the Conservative Party in 1935 — no matter apparently about the interests of the country, or the danger we ran in the process. The interests of the Conservative Party come first. And, in fact, we ran pretty close to the danger of a war with Italy, in a state of unpreparedness in consequence. Behold the patriotism, behold what the country owes to the sense of responsibility of Mr. Baldwin!

But why hasn't the Labour Party attacked on all this? They ought to have attacked the National Government, not for rearming, but for the criminal irresponsibility of delaying Rearmament until it was two years too late. The party has failed to capture the patriotic cry, it has failed to identify the interests of the country with itself — which is the one thing necessary if it is not only to win but retain power. Moreover, it happens to be the truth: if I did not know that it is the Labour Movement that expresses the interests of the great bulk of the people of this country, I should not be a member of it. It is wonderful how a comparatively small number of people, just by their possession of economic power, can put it across that *they* are the real custodians of the nation's interests, that

---

[1] "I had the greatest difficulty in preserving the use of the bombing aeroplane."— Lord Londonderry.

theirs is the national party, while the Labour Party is — a class party. That is, to put it quite simply, a lie; but it is very effective. This matter is one of extreme importance. *Until the Labour Party captures the position of being the real exponents and defenders of the nation's interests from the Conservatives, it will never be sure of power, even if it wins an election.* The October Revolution in Russia was never safe until it became identified with the national interest and defence of Russia. It was the same with the French Revolution, which was never strong until the remnants of the old régime led the attack on the sacred soil: then the Revolution became identified with France. It is a well-known historical law. But then the one thing that history teaches is that men never learn from history. Each time they have to learn it all over again in their own bloody experience: the incredible fools!

Now: if this country is led into a first-class political defeat, say over colonies, or in the Mediterranean, or possibly even in the Far East, because of the confusion and ineptitude of the policy pursued since 1931, there will be a great opportunity for the Labour Party to identify itself with the interests of the country. It should be clear at last. And that is the *sine qua non* of continuous Labour rule in this country. But will the party seize its opportunity? Does it see the possibilities of getting a hold on power at last here? Not if one may judge from the speeches coming from the floor of the Conference at Bournemouth; not if one may judge from the hopeless doctrinaires, the illusionists, the crazy unrealists, who are the despair of the party. It would be invidious to mention names; besides, there are so many, and of these quite a number are Members of Parliament.

But the party is beginning, in its leadership, to take the right line at last about all this. The Trade Unions have been right all along, as they almost always are: they are practical men, who know what the world is really like. When one thinks of the intellectuals of the Labour Party, who are almost always wrong, one thanks God for the Trade Unionists. Mr. Attlee's speech in the debate on the Address entirely failed to point out how British interests were being sold by the upper

The Prospects of the Labour Party (1937)

classes, and to identify their defence with the Labour Party. He even gave the case away by attacking the Government for its "cynical" alacrity to defend British shipping in the Mediterranean, while utterly failing to defend the interests of collective security. This is a fatal line; it enabled Mr. Chamberlain brazenly to take Mr. Attlee to task for disregarding his country's interests, warning him of the consequences if the people have to regard the Labour Party as responsible for their not being defended. That — from these people who have sold the interests of their country from Shanghai to Spain, in the interests of their class and their friends abroad! That should show the silliness of taking such a line. But the attitude of the League fanatics is no less hopeless, as Mr. Eden was able to show in relation to Mr. Noel-Baker, to whose speech he gave pride of place for unreality. What indeed is the point of saying that the Spanish question should be referred to the League, when the League members are divided from top to bottom on the question? Mr. Noel-Baker seems to be going on with his head in the clouds of 1929–1931, and his feet never on the ground. That all came to an end with Hitler.

It remained for Mr. Morrison, in his speech at the close of the debate, to give the Labour Party the only possible lead, for which the circumstances are crying out: to lay the responsibility home to the National Government for the dangerous crack-up in our foreign situation, the retreat we have made all along the line because our upper classes, half of them, have been sabotaging on the side of the Japanese, Mussolini, and Hitler. And now, if it is too late, and we have to think wholly in terms of air-raid precautions for our towns, the responsibility is theirs. We are not opposed to armaments for the defence of the country, but this is what happens when the upper classes of the world win, and the working-class movements are shattered because they believe in peace. Nevertheless, their victory in the end is the only hope of peace in all the world. *That* is our line; that is what we know to be true, and all the evidences of the contemporary world go to prove it.

When it came to air-raid precautions, it is interesting, too,

The End of an Epoch

that it was Mr. Morrison who put his finger on the cause of our belatedness and unpreparedness: Sir John Simon's tenure of the Home Office from 1935 to 1937, during which no communications were established between the central Government and the local authorities on this most pressing subject. It is not surprising, though it may have been forgotten, that Sir John Simon was also Foreign Secretary from 1931 to 1935.

It is clear from the new session in Parliament, as it has been to my mind for the last four or five years, who the real leader of the Labour Party is. One has only to have attended the Labour Conferences of these years to recognise that Mr. Morrison's is the authentic voice of the party. There were special circumstances that prevented his election to the leadership in 1935. But the conviction is growing, and not a moment too soon, that Mr. Morrison must become the leader of the party if it is to win power.

There is no man in the Labour Movement who commands such general confidence. The results of the municipal elections have made that abundantly clear; though the advance in most parts of the country was not very considerable, in London it was astonishing. There are now 17 boroughs out of 28 under Labour control in a city one had never expected to see under Labour rule, even when the party was entrenched in Westminster. This is a vote of confidence in Mr. Morrison; for his personality and leadership are the chief assets that Labour has in London. It is obvious that the lines on which Morrison runs the London Labour Party need to be extended over the whole country. It would mean a new spirit in the organisation; it would mean a new inspiration to Transport House. Most important, it would give what the modern electorate needs above all — leadership. The masses do not understand programmes or policies — and there has been so much waste of time in the Labour Movement wrangling over programmes and policies. Hardly any of it matters; what the people understand is a man.

Lastly, is it not necessary that the party should do everything to extend its basis, to broaden its appeal to the country,

The Prospects of the Labour Party (1937)

as a condition of power? I think it is clear that it should. The fact is that though the Labour Party may, and does, represent the interests of seven-tenths of the nation — even if they are not aware of it — yet it does not *look* so much like a national party as the Conservatives do. I have said above that its *personnel* needs strengthening; though we have the support of the young, we need the ability and experience of more men in the fifties and forties. If we are to command the support of the nation, in these critical days for parliamentary and democratic régimes, and not merely be tolerated *pour faute de mieux* as in 1924 and 1929, we need to recruit to ourselves and make use of the ability and prestige of all progressive-minded people who will work with the party. The names of half a dozen such men, of high ability, who would add strength and confidence to any Labour administration, will leap to mind: it would perhaps be invidious to name them.

The key to this broadening of the appeal of the party lies in an understanding with the Liberals, and nowhere else. At Bournemouth the Left Wing was reconciled and brought into step with the party. If it is to win a majority, it must now extend its appeal towards the Centre. A great deal of time and energy has been wasted on the question of a United Front with the I.L.P. and the Communists. The game is not worth the candle. Each of these little groups consists of no more than a handful of voters, whose nuisance value is out of all proportion to the services they can render; and their nuisance value would be correspondingly greater in association than without. Anyone who followed Mr. Morrison's reply to the Unity Campaigners at Bournemouth will recognise that he left their case with not a leg to stand on. He subjected it to three fair and strict tests: Would such a United Front damage Fascism? Would it hurt the Tory Party? Would it help the Labour Party? He showed, beyond possibility of answer, that in each case it would not.

The I.L.P., consisting of a group of fissiparous sectarians, are simply *pour rire*. The Communists are a more serious matter; nor is their present line of policy, now that Moscow is making them eat the dirt of the past twenty years, so objectionable.[1] But they should have thought of their present moderation twenty, or even ten, years ago; before their policy of aiming at the destruction of social democracy had fulfilled its purpose in Germany, with what consequences we can all see. The best contribution the Communist Party could make to the cause they have so much at heart, would be to cease to exist. Moreover, it is the business of all working-class organisations to take their place within the Labour Movement in any case.

The next step necessary to make sure of power, other circumstances being favourable, is an understanding with the Liberals. There were as many as a million Liberal voters as late as 1935; the greater portion of the reactionary Liberals are now voting with the National Government. It may be assumed that the majority of those who remain are progressives, prepared to support a policy of the extension of public control over industry and finance, no less far-going than that of the *Liberal Yellow Book* of 1929, and a whole-hearted foreign policy of collective security. It should be remarked that in Denmark, in these last nightmare years of reaction and relapse into barbarism, a Social Democratic government has maintained control with the aid of the progressives represented in the Cabinet. Similarly in Sweden, the Social Democrats have chosen to govern in association with and resting upon the support of the Farmers' group, rather than with the Communists. That has given a period of stable government to Sweden, remarkable in all Europe for its works of social progress and leadership in economic technique, the control of the trade cycle and unemployment. The Scandinavian models are those we should have in mind rather than Russia on the one hand, or Germany on the other. I am not impressed by Mr. Attlee's rejection of an understanding with the Liberals; and there are certain signs that the party is moving away from that too rigid orthodoxy. Mr. Attlee says

---

[1] Of course this changed overnight with the Soviet-German Pact of 1939. The war that followed as a natural consequence was an " imperialist " war on the part of England and France, and to be obstructed, not aided, until the day when Hitler invaded the soil of Holy Russia.

The Prospects of the Labour Party (1937)

that the fact that Liberals are not Socialists makes it hard to work with them.[1] But would it not be excessively difficult to find anybody with whom one could work if Socialism were made a precedent condition? Are most Labour people any more Socialists than Liberals are?

To come down to practical politics: the Labour Party is committed now to the extension of public control in industry and banking. I imagine that most progressive Liberals who remain would be willing to cooperate in some such agreed programme. The model of public control, the " public corporation ", to which the Labour Party is committed, is after all a Liberal invention. The least the Labour Party can do is to find out what possibility there is of an agreement on these lines. Full cooperation in the country can only follow an understanding arrived at at the centre. I note that Mr. Attlee does not " rule out such a thing as an impossibility in the event of the imminence of a world crisis ". In these days we are never far removed from crisis; and it may be that with such a crisis upon us, we shall draw together after all. That would provide the basis for a period of strong and progressive government such as this country needs, and for the kind of lead from us which the world has awaited in vain since 1931.[2]

[1] *The Labour Party in Perspective*, p. 132.

[2] Alas, it never came — so great is human stupidity and folly — until the very gravest danger to our existence in our history as a nation forced all parties together to meet it. Why hadn't they the sense to come together beforehand to prevent it — so, perhaps, they would have saved millions of simple people's lives?

X

WHAT IS WRONG WITH THE CIVIL SERVICE?

[1942]

WHAT is wrong with the Civil Service? is a question that has been agitating the minds of those in touch with Government and concerned with public affairs for some time. It would be even more to the point if they had got agitated rather earlier and done something about it sooner, before the war came down upon us. For now with a war on our hands which faces us with the greatest test in our history, we find ourselves with a Civil Service — after all, the main instrument of Government — that has considerably deteriorated since the last war, and has shown itself in late years, at least in its upper reaches, not up to its job. Sir William Beveridge, himself one of the great civil servants of his generation, tells us in a letter to *The Times* that in some ways departments were less well organised for war in 1939 than they were in 1914, while everybody in the know is aware of the decline in ability and aggressive energy at the top. Another great civil servant of former days described to me the deplorable effect the rule of yes-men has had over the past two decades, the age of Baldwin and Chamberlain — " who appointed yes-men, who appointed other yes-men, who appointed other yes-men down to the third generation — so that never have I come across such a set of men who had no idea of getting anything done, and who wouldn't know how to get anything done, even if they had wanted to ".

Since the war began, it is only fair to add, some steps have been taken to remedy this state of affairs. There has been inevitably a great expansion in the Civil Service; new men of great ability have been recruited for the duration of the war.

What is Wrong with the Civil Service?

But I do not know that there have been changes at the very top. In particular, the equivocal figure of Sir Horace Wilson — the partner of Neville Chamberlain in the deplorable policies of those last years which made the war inevitable, encouraged our enemies, and lost us our friends — that man is still at the head of the Treasury, the most important post in the whole Civil Service, with a larger hand in making appointments than anybody else. I cannot understand why he is still there.[1]

It is the men at the top who matter. And at last we have a most revealing book in which we, who are outside and know nothing of what goes on in the upper reaches of Government, can study the character and composition and the outlook of the extraordinarily small body of men who are quite unknown to the public and yet govern us from behind the scenes.[2] Of the half-million or so members of the Civil Service in peacetime, it is only a tiny fraction who count in the formation of policy: Mr. Dale estimates some 550 in all. These form the Higher Civil Service, which he investigates for our benefit. And out of that number there are no more than twenty to thirty men who exercise a decisive influence in forming, along with Ministers, the lines of policy and executing it. A sort of secret Cabinet behind the shifting kaleidoscope of the public one, the well-known array of public faces that come and go; with this difference that the high permanent officials may say, like Tennyson's brook, " But we go on for ever ".

The usual pretence is that the civil servant has nothing to do with the framing of policy; it is his business merely to execute it. It is the sort of political cliché one may expect from a Horace Wilson. Not so Mr. Dale, who tells us explicitly and honestly: " The most important function of the great permanent official is not to carry out decisions already taken by Ministers, but to advise them what decisions they should take ". And he goes on to say, what is obvious, that " when the question at issue is a large question of policy not to be deter-

---

[1] He went shortly after, and was replaced by a man of first-class ability, Sir Edward Bridges: a key-man among the small group of administrators to whom we owe the marvellous efficiency of the conduct of the war.

[2] H. E. Dale, *The Higher Civil Service of Great Britain*. Oxford University Press.

mined on facts and figures alone, the opinion even of a civil servant is not the product of an abstract intellect, but of a whole personality ".

That being so, the social background and outlook of this handful of men who govern the country behind the scenes is a factor of the greatest importance in its political life. Mr. Dale gives us a revealing analysis of their social origins and education, their daily life, and what may be inferred of their outlook. It confirms what one had thought already: that these men belong almost wholly to the upper middle class and incarnate in themselves the virtues (and the shortcomings) of that stratum of society. They are almost all public school and university men. Of eighty of the chiefs of the Civil Service, only two could be said to have started with an elementary school education.

One sees the Higher Civil Service then as a stronghold of the middle class. No wonder the great permanent officials find it easier to get on with a Conservative Government. (Mr. Dale: " For reasons to which I have already referred, it is probable that the state of beatitude is more often attained when a Conservative or National Government is in power ".) And when we reflect that this country has been governed almost wholly for the past twenty years by a " Conservative or National Government ", then the Higher Civil Service has enjoyed a long period of beatitude. It cannot escape attention, when we consider who is responsible for the mess that this country has been landed in, that these 550 elderly, respectable upper middle-class gentlemen, the Higher Civil Service, have had their share in it. I do not know that they sounded any alarms, even discreet ones, about the course we were going on. No wonder Mr. Chamberlain was such a favourite with them; he was one of them, absolutely of them. With all their respectability, their conscientiousness, their laboriousness, their devotion to duty, we cannot but recognise on the other side a certain timorousness, a collective fear based on a feeling of social insecurity, unimaginativeness, a blinkered devotion to routine without considering the great decisions that a country must make if it is to live in times of danger;

What is Wrong with the Civil Service?

in the end, an absence of that high courage that aristocrats possess, and adventurers are not wholly without. Chamberlain was their ideal, their favourite, be it noted; not Lloyd George or Mr. Churchill.

What is wrong, then, with the Civil Service is something that is wrong with the middle class and its nineteenth-century standards, in this period of its decadence and decline.

To this Mr. Dale adds an indictment of his own; it is obvious if you look between the lines of his chaste pages that he is disquieted. The Civil Service mind is one that is antithetical to action. Upright, fair, impartial (so far as it can be), it does not reach out to take responsibility and power — as bureaucrats are constantly accused of doing. Its instinct is rather to evade responsibility, to pass it on somewhere else, to shuffle it on to a committee, that irresistible resource in the complicated game of battledore and shuttlecock that goes on between Parliament and Civil Service in a parliamentary democracy. Mr. Dale is quite clear that the responsibility for decisions must be made to sit fair and square upon the shoulders of some one person. In fact, we want more people of the calibre and type of Beveridge, Arthur Salter, E. F. Wise, Sir Robert Morant, and fewer of the Horace Wilsons; and it is noteworthy that all the first four, great men in their way, left the Service, while Sir Horace remained and got to the top. There must be something wrong.

There is a chapter on " The Official Temper of Mind and Disposition " that is very revealing. The whole psychological pressure arising from their work is towards conciliation, compromise, collective conformity. It seems to me that they pay far too much attention to the complaints and difficulties, often trivial ones, that Members of Parliament spend so much of their time handing on from their constituencies. Any kind of parliamentary business has first claim upon the civil servant's time; and far too much consideration is given normally to circumventing difficulties and objections that are hardly worth considering seriously — especially when what is wanted is more leadership, more ruthlessness, more decision. All that attitude arises partly from the nature of their work and partly from their creed. Mr. Dale has an illuminating summing-up of the Civil Service creed, of which a foremost article is that " There is much to be said on both sides of the question ", and no doubt they proceed to say it in elaborate memoranda in the long-winded hieratic language, so non-committal and nicely evasive, which Mr. Churchill asked the Service to discard for the duration of the war. The truth is that very often there is *not* much to be said on both sides of the question, and what is wanted is a clear decision cutting the cackle and personal responsibility for it. The Civil Service creed seems to be that of a disillusioned liberalism, a deadening, if not a dead creed, sucking the vitality of those who continue to hold it in a world that belies its standards and very assumptions. (Mr. Dale, oddly enough, calls it Realism ; but Realism means recognising the truth without being paralysed by it. And he himself pays tribute to the intellectual ascendancy of a dead creed over the minds of elderly civil servants when he says that " If the Liberal Party had the luck to find a man of genius — a Disraeli or a Gladstone — to be its leader, in ten years or less it might be in power again ". Anybody who can believe that simply doesn't understand the real facts and forces that operate in politics. The whole social foundations of Liberalism have collapsed in the contemporary world. No amount of genius can restore life to a corpse.)

All this relates rather to what is wrong in general with the Civil Service. But what is it that is wrong in particular at this present testing time ?

There has in recent years been an enormous, an almost unbearable increase in the work of administration ; the higher officials are shockingly over-worked ; hardly any of them has any time to think ahead. All the more reason to cut out whatever may occupy the time of important officials superfluously in correspondence or parliamentary questions. Mr. Dale is of the opinion that a certain proportion of these officials show the effects of having over-worked when young to get their footing in the Civil Service, and in middle age are prematurely tired. Too much importance has been attached to the examination system as a mode of entry. Mr. Dale says that, of the four outstanding men he has known in the Service, men remarkable for their energy, driving force, and public zeal, not one " entered the Service by the ordinary portal of the Administrative Class examination ". That gives food for thought when we come to consider remedies. Sir Robert Morant, by universal acclaim the greatest civil servant of his time, a second creator of the Board of Education, and creator of the Ministry of Health, was a late recruit to the Service and did not enter by way of examination. And what a contrast that great man, with his iron resolution and single-minded determination, his constructive vision and tireless energy, was to the high official who expressed a heartfelt admiration for a distinguished Secretary of State on the ground that he could " always discover a reason for doing nothing that had not occurred to his permanent staff ".

That kind of mentality is responsible for our being where we are today. And it goes without saying that a totally different mentality is what is required of the Civil Service in the crisis of the war. Yet it is at this very time, Mr. Dale tells us, that the effect of the last war upon the personnel of the Civil Service, in loss of men, in wastage due to physical overstrain, etc., is at its maximum. There is further the age question. Anything under sixty, he tells us, is a moderate age for a modern Cabinet Minister. It is disgraceful, when one considers that Pitt and Fox were Secretaries of State at twenty-three, and things went far better with us then than they do today. The fact that our politicians are too old is one of the prime causes of our decadence as a nation. And, of course, the highest permanent officials keep pace with their opposite numbers of the public Cabinet; they are all between fifty and sixty, usually nearer sixty than not.

But a further factor peculiar to the Civil Service has been responsible for a marked decline in initiative, in independence of mind, and willingness to assume responsibility: the extension of Treasury control over the other departments. Since the last war the Treasury, as a sort of establishment office having control over the purse-strings of the other departments, has used its position to extend its control over personnel and appointments throughout the rest of the Service; and the

The End of an Epoch

Head of the Treasury has come to occupy a quite unconstitutional, but enormously influential position, as virtual Head of the Civil Service. In every other department the Treasury has its little gang of fifth-columnists: a kind of dead hand so far as departmental initiative and independence are concerned, by no means dead where appointments are in question. As for remedies, the first thing to do is to remove the dead hand of the Treasury upon the other departments, and restore them their independence. I would add to that, removing the present Head of the Treasury, that permanent reminder of the humiliations of the Munich period.

For the rest, it is fairly clear from the above diagnosis of what is wrong, what our remedies should be. I began by saying that a great many new men of ability and drive have come into the Service from outside for the duration of the war. They should be given the posts their abilities merit without any question of seniority. The younger they are the better. And it is to be hoped that the State will be able to retain the services of some of them at least after the war. Let them take the places of the elderly men who are too tired and not up to their jobs. The post-war period will be a time of difficult adjustments and reconstruction not less arduous for the Service than the war itself. It is to be hoped that the opportunity will be taken to bring in men of ability and drive from outside, by appointment, and by-passing the routineers of the past two decades. During that discouraging period there was a notable drift of some of the ablest men away from the Service, into business or service abroad. I hardly think that business will have the same attractions after this war as it had after the last. So there may be the men available, if they are properly made use of, to belie Mr. Dale's depressing prognostication: " If we can judge from the experience of the last war, it is probable that when the tempest is over the permanent Civil Service will emerge from the waters not fundamentally altered from what it was before the floods covered it ". In my view it will need new men at the top, and a new spirit infused throughout it, if it is to take its proper part in the gigantic work of reconstruction awaiting it.

XI

THE DILEMMA OF CHURCH AND STATE

[1936]

"It is a fundamental principle that the church, that is, the Bishops, together with the Clergy and Laity, must in the last resort, when its mind has been fully ascertained, retain its inalienable right, in loyalty to our Lord and Saviour Jesus Christ, to formulate its faith in Him and to arrange the expression of that Holy Faith in its form of worship."—Archbishop Davidson's declaration on behalf of the Bishops, July 2, 1928.

"The difficulty has always been, and still is, to reconcile their claims to spiritual independence with an Establishment over which the authority of the Crown is, in the last resort, supreme."—*The Times*, on the Bishop of Lincoln's case, April 1889.

THESE two quotations indicate very precisely the nature of the dilemma that has underlain the relations between Church and State in the past fifty years or so : ever since, in fact — as the Bishop of Durham as an historian has been quick to see, or at least quicker than others to express — the social and political conditions that made the Establishment a reality disappeared.[1] Now, with the disappearance of those conditions, as he has recently expressed it in the debate in the Church Assembly on the Report of the Archbishops' Commission on Church and State, "the Church of England lingers on the scene as practically the solitary survivor of the Established Church, a link with something that was once universal, but in modern conditions has been found impossible in country after country". It may indeed be thought remarkable that the Establishment should have survived the shocks and conflicts of this century, in which in one country after the other, beginning with France in 1906 and going on to the Weimar

[1] Cf. his article, "Ought the Establishment to be Maintained?" *Political Quarterly*, September 1930 ; and his *Disestablishment*, 1929.

Republic in Germany and Spain in 1931, the special relation of one Church to the State has been brought to an end. It is not only to the extraordinary institutional conservatism of this country that this is due, but in a special sense to the general damping-down of all such latent issues that has come about with the strengthening of Conservatism after the War of 1914–18, and more particularly after 1924. *Quieta non movere* is very much the motto of this period we have almost lived through: the Baldwinian epoch as it will be regarded by the future historian.

Nevertheless, that it is not possible to damp down the issue to extinction with general forbearance and good-will is evident. True, we have had nothing like the bitter conflict with the Church such as France has had to go through, nor such as Spain and Germany are now experiencing in their respective forms. Yet in a number of ways the latent issue bursts through the complacent surface, the quieter circumstances of this country's politics. The tithe question is a serious one in some districts: it has even necessitated a Royal Commission. The disputes over the Revised Prayer Book occupied the forefront of public attention in 1927 and 1928. There is a party conflict in being within the Church itself which from time to time breaks out overtly into politics, though the division is not on party lines in Parliament. The appointment by the Archbishops of this Commission to go into the question of the relations between Church and State was directly due to the rejection of the Revised Prayer Book by Parliament, and was motivated by the desire to find some way out of the dilemma of which that rejection was a forceful and startling reminder.[1]

The Bishops had not the courage to take the responsibility for going into the *whole* question of Church and State; the Commission when appointed was directed " to inquire into the present relations of church and state, and particularly how far the principle, stated above (*i.e.* by Archbishop Davidson),

---

[1] *Church and State*: Report of the Archbishops' Commission on the Relations between Church and State. Vol. I: Report; Vol. II: Evidence. (The Church Assembly.)

The Dilemma of Church and State

is able to receive effective application *in present circumstances* in the Church of England, and what legal and constitutional changes, if any, are needed in order to maintain or to secure its effective application ". That is to say, the Commission is chiefly concerned with how to improve the position from the point of view of the Church, on the basis of the existing Establishment. That fact in itself was enough to disinterest from its proceedings a number of churchmen, headed by the Bishop of Durham, who feel that there is no remedy to be obtained within the Establishment. As Dr. Henson says in his letter to the Commission :

If I seemed to allow that an adequate reform of the existing Establishment is really within the sphere of practical politics, I should be gravely misleading English churchmen. In the circumstances of our modern world, I do not think that the maintenance of the Establishment is a legitimate object of Anglican effort. The resolution of the Church Assembly which states the purpose of the Commission and provides its " reference " assumes that the present relations of church and state may be found on inquiry to be consistent with the autonomy of the Church of England as a spiritual society; and that, even if not, they may be made so by some " legal and constitutional changes ". The first assumption appears to me to be apparently and confessedly false : the second, to be so remote from practical value as to be deeply mischievous.

Allowing for the exaggerations of ecclesiastical rhetoric, it is evident how deep the rift is here; and that the findings of the Commission can have no value for the large and growing body of opinion within the Church which looks to Disestablishment to free its life as " a spiritual society ".

This body of opinion, which was in any case strong among Anglo-Catholics, received a large accession of strength among other sections of Church opinion as the result of the Prayer Book rejection. To them it was a signal reminder that the Church was not free to order its spiritual and doctrinal life as it thought good; that Parliament was legally and constitutionally the ultimate authority in these matters no less than over the Church's property, and that it was intolerable that the forms of service upon which the highest authorities in the

The End of an Epoch

Church had been at work for twenty years, that had been passed by overwhelming majorities in the House of Bishops and in the House of Clergy, and by large majorities in the House of Laity and by practically all the Diocesan Conferences, should be rejected by a House of Commons in which Jews, unbelievers, Nonconformists, and Catholics — and a solitary Parsee, to the great scandal of the devout Lord Birkenhead — voted equally with Anglicans.

Had they acted on strictly secular grounds [says Lord Hugh Cecil], they would not have exceeded their rightful jurisdiction. But the rejection was mainly prompted by purely religious considerations, relating to the doctrine of the Eucharist and to changes in the service for celebrating it, and to the permission of the reservation of the consecrated Sacrament. These were purely spiritual matters which the church had jurisdiction from Christ to order, but the state had none.[1]

It is not for us here to point out the historical inaccuracy of Lord Hugh Cecil's position, though the fact is that the religious position of the Church of England was as much determined by the secular authority as its forms and externals. It was laid down by Henry VIII and Elizabeth; such as it was made by them, the Church of England has substantially remained — and the Cecils had a considerable hand in shaping it. It might be added that history affords no backing whatever for the High Church party's reading of the Church of England's past; it completely substantiates *per contra* both the Low Church and the Roman view that the Church of England was shaped and determined by secular authority.[2]

---

[1] *Report.* Vol. II, Evidence, p. 3.

[2] Cf. Sir Lewis Dibdin, *Establishment in England*, pp. 4-5: "How entirely the hierarchy were overruled and the Christian Laity of England asserted itself is best shown by what happened in the first year of Elizabeth. The Canterbury Lower Convocation, truly representing the mind of the clergy, passed resolutions in favour of the Mass and the Pope's jurisdiction, and declaring that laymen could not properly meddle with questions of faith, sacraments or ecclesiastical discipline, which were for the clergy. These resolutions were taken to the bishops, and Bonner presented the Resolutions to the Lord Keeper, by whom they were graciously received and then laid aside. Meanwhile Parliament passed the two Acts 1 Eliz. ch. 1 (Supremacy) and 1 Eliz. ch. 2 (Uniformity)." To quote Professor Gwatkin: "The Acts of Supremacy and Uniformity and the injunctions form the transition

The Dilemma of Church and State

The rejection of the new Prayer Book upon which the hopes of the restoration of order within the Church were based, left the position still more confused. All clergymen at their ordination swear that they will abide by the Prayer Book of 1662, and all bishops at their consecration that they will administer it. Yet it is safe to say that there is hardly a church in England where the Prayer Book is adhered to without some deviation, either by way of addition or subtraction; while for the bishops to administer such a situation according to the letter of the law is manifestly impossible. A strict interpretation of the law would undoubtedly say that the whole clergy of the Church of England were guilty of breaking their oaths of obedience — far more clearly than the clergy under Henry VIII were guilty in accepting Wolsey's legatine authority, for which they were fined enormous sums and forced to make submission, the step by which the Reformation was set in train.[1] The Commission, while recognising the rigidity of the law of public worship applying to the Church, describes the situation in grave terms:

No one obeys the law so construed. Not the clergy, since there is scarcely one of them who makes no change in the authorised forms of service. Not the bishops, who are charged to see that this impossible law is carried out. Worse still, by the Declaration, as interpreted by the Courts, every priest solemnly undertakes to do that which in fact none of them actually perform. No wonder discipline has suffered. No wonder the laity are uneasy. It is difficult to find temperate words to apply to such a state of things. The situation can only be described as deeply insincere.[2]

We question how far, grammatically, a " situation " may be described as " insincere ".

As again, when the Commission particularises: " To accept an office on certain terms and then to refuse to carry out those terms, but nevertheless to retain the office, shocks from the old law to the new; and they were all the work of the Laity. Convocation was not even coerced into formal acquiescence, as in Henry's time, but simply ignored." In other words, the Reformation was accomplished, and the Church of England created, against the representative will of the clergy by the State.

[1] Cf. J. R. Tanner, *Tudor Constitutional Documents*, pp. 16-20.
[2] *Report.* Vol. I, p. 85.

the public conscience, and it seems clear that a situation which drives such men to such action is indefensible ", a more exact attention to logic would seem to indicate that it is the *action* which is indefensible, and not the situation. And that is exactly what the Protestant party in the Church concludes, as Sir Thomas Inskip in his evidence says bluntly:

> I do not understand the position of those who complain of the fetters of the state and yet depend on endowments which rightly or wrongly belong to the Church of England by virtue of parliamentary and political action. The church cannot face both ways, as its bishops and clergy — generally speaking — appear to be doing at present.[1]

In fact, if it were not for the forbearance of the State, its tacit connivance at the breach of the law by the whole Church of England since the rejection of 1928, the position could not continue. There certainly was reason enough for the appointment of the Commission; whether its recommendations meet the situation or provide a remedy is what we have to inquire.

The present position is, broadly, that after the rejection, the bishops met and determined to continue as if the Revised Prayer Book had been passed. More exactly, after hearing Archbishop Davidson's solemn declaration of the spiritual independence of the Church, the bishops proposed resolutions to the Convocations stating that they felt bound not to interfere with clergy whose deviations from the Prayer Book were within the limits which the Revised Prayer Book would have sanctioned. As the result of the bishops' protection, the Revised Prayer Book is in operation in practically every diocese in the country — but without statutory sanction, so

[1] And for the bishops, Sir Thomas Inskip says: " All the bishops on the Bench, and very many of the Cathedral Clergy, have accepted their commissions from the state. I do not suggest for a moment that they did not believe the state to be divinely guided in these appointments, but I cannot for myself see the distinction between, on the one hand, asking Parliament to express its views on the Book of Common Prayer, so as not to alter the Book without Parliament's concurrence, and on the other hand, accepting the Prime Minister or the Lord Chancellor as the authority to appoint the bishops. I know one bishop who was an ardent critic of the action of the House of Commons, but who, on more than one occasion, solicited the help of his Member of Parliament in persuading the Prime Minister to make him a bishop." (*Report*. Vol. II, Evidence, p. 102.)

The Dilemma of Church and State

that the bishops' own position is extremely insecure. Nevertheless, it is only fair to admit that the bishops have secured a far greater measure of obedience on this basis than might have been expected — even if the basis is an illegal one. This in itself is a pointer to where the solution really lies — to the day when the bishops will be the ultimate authority responsible for order in the Church, and not Parliament. Still the position remains an illegal one, and, as the Report says:

> So far from solving the problem, it conspicuously illustrated a legal, and indeed a moral, situation with which it neither was nor is possible or right to rest content. For even apart from the legal obligations imposed by the Acts of Uniformity, every bishop, priest, and deacon, at his admission to his Order, and to any benefice or legally recognised position in which to exercise the functions of that Order, solemnly declares that in public prayer and administration of the Sacraments he will use the form in the Book of Common Prayer prescribed and none other, except so far as shall be ordered by lawful authority.[1]

The mere recommendations of the bishops without the sanction of Parliament, so long as the Church remains established, are not lawful authority.

The Report begins with an historical introduction leading up to the present state of the controversy between Church and State, of which one can only say that it well illustrates the impossibility of history being written by a committee, for it is both confused and, in a curious way, ultramontane. It does not so much matter that the Commission should say, " The distinguishing mark of the Church of England has always been its close connection with the territorial state " — though in fact it is not more true of the Middle Ages in England than, say, in France or Spain. Nor does it much matter that there is a tendency in the Report to write the history of the medieval Church in England backwards from the modern State Church independent of Rome; to suggest the independence of the English Church in the Middle Ages, almost as if Maitland had never written. The Report omits to notice the *Suprema Potestas* of the Pope, and the fact that the English Church was rather

[1] *Report.* Vol. I, pp. 39-40.

more under the wing of the Pope than in most parts; as it omits also to mention that Henry VIII did interfere in doctrinal matters — he virtually determined what the doctrinal position of the State Church was, and what it was not — and received the obsequious thanks of the clergy for his rare theological wisdom. Again, there is no point in the Report saying that, under the Tudors, " There was no abandonment by the church of its claim to spiritual independence. In strict theory, Parliament had no concern with matters of doctrine or ritual." This is merely disingenuous; for, in fact, Elizabeth and Parliament did determine the issue of doctrine and ritual in 1559, with her the Royal Supremacy was continuously active and Parliament is the heir to the powers of the Royal Supremacy. These are historical matters, and they affect the perspective in which the whole issue is seen.

But it may be of greater practical importance to observe the totalitarian character of the claims put forward for the Church by these eminent Commissioners: it is in the best ultramontane tradition.

Christian faith claims to control, after its own manner, the whole of life. And vital Christianity refuses to be regarded as a department of civilised society. If only it could consider religion, in the popular phrase, as " a private affair between a man and his Maker ", the church could acquiesce in regarding itself, and being regarded, as a devotional club which those who are so disposed are at liberty to join. But it cannot acquiesce in such a view of its nature and function without repudiating its trust. For it believes that it is more than a voluntary society. It is the Body of Christ, the organ of the will of the divine Lord.[1]

Which body? we may well ask; for it is noticeable that each particular sect regards itself as " the Church ", and refers to itself as such. In the body of this Report and in the evidence, the Church of England is described as " the Church "; the Church of Scotland is also " the Church "; indeed in the Articles to which the Church of Scotland Act, 1921, gave statutory authority it describes itself as " part of the Holy Catholic and Universal Church ". No doubt the Wesleyans

[1] *Report.* Vol. I, p. 7.

The Dilemma of Church and State

and the Particular Baptists also regard themselves as " the Church ". But it is regrettable to observe that the one body which has a historic claim to speak for Christianity, the Catholic Church, recognises none of these claims. Yet, the Report asserts, " the Church of England continued, as it still continues, to represent the Christian faith of the nation as a whole ", a claim that, it is safe to say, neither Catholics nor Nonconformists would allow. It is not to be wondered at that the Report should state, " the distinction between secular and spiritual interests in the life of the community becomes also increasingly unreal ", though Dr. Henson, a more acute observer of the signs of the times, could tell them that the whole tendency is the opposite, towards the delimiting and disentangling of Church matters from the State. The Commission's recommendations are themselves evidence of the trend. All we need say, in conclusion on this point, is that if this high-flying line represents the attitude of the Church, it is difficult to see how a direct collision with the State is avoidable, as direct in this country as in France, or Spain, or Germany. For there is no solution on the lines laid down by the Commission. " There is no department of the common life of the citizens of a community into which both church and state do not claim to penetrate, and which they do not seek to direct or influence." The conflict is " inherent in the very nature of church and state ", they say. Of course it is — on their conception of the nature of Church and State. They reject the conception of the Church as a voluntary society, contemptuously, as " a devotional club ". But all Churches must be regarded as voluntary societies by the State, if only for the simple reason that the claims of religious sects are divided and mutually conflict. (We do not need to state other reasons, such as that the claims themselves are untrue.) It is the State's function to stand neutral amid the clamour of them all; and therefore is it said by the political philosopher, " It is an incongruity on the part of the state to endow one of several religions professed by its citizens, still more to identify itself with such a religion ".[1]

[1] R. M. MacIver, *The Modern State*, p. 20.

The End of an Epoch

So much for the political philosophy underlying the Report. To come to its proposals: these may best be described as an attempt to carry the degree of freedom as regards the legislation of the Church which was attained by the Enabling Act (1919) a stage further. It is the freeing of the Church's legislative process, so that the measures it agrees upon may be carried into law, that is the crux; the rest, discipline and the reform of the ecclesiastical courts, are subsidiary, in a sense even, consequential. The freedom that was conferred by the Enabling Act was only a limited freedom. Briefly, it enables the measures of the Church Assembly, after coming before the Ecclesiastical Committee consisting of 15 members of the House of Lords and 15 of the Commons, to be laid before Parliament; and upon each House passing a resolution that the measure shall be presented for the Royal Assent, it thereupon has the force of an Act of Parliament. That is to say, the parliamentary process is short-circuited for Church measures; they may be rejected, but not amended. On the whole, for the usual routine measures of the Church, this process has worked well; on the other hand, on the Revised Prayer Book, whose passage it was designed to facilitate, it broke down: the measure was rejected by a majority in the Commons.[1] If the measure could have been amended, it would doubtless have passed; but in that case it would never, without the permission for Reservation and the new Order for Holy Communion, have been agreed to by the Anglo-Catholic party in the Church. Such is the dilemma.

The Report now proposes a distinction between ordinary administrative measures of the Church, which should go through the present procedure under the Enabling Act, and spiritual or doctrinal measures dealing with the formularies and ritual of the Church, for which they propose a special procedure. They propose that a Committee of the two Archbishops, the Lord Chancellor, and the Speaker should grant a certificate to measures which " relate substantially to the spiritual concerns of the Church of England ", and certify-

[1] For the working of the Enabling Act cf. Lord Hugh Cecil, *The Church and the Realm* (1932).

ing that "any civil or secular interests affected thereby may be regarded as negligible"; upon the granting of such certificate, such measures, provided that they have passed the three Houses of the Church Assembly separately and been twice approved by the Diocesan Conferences of three-quarters of the English dioceses, may go direct for the Royal Assent and pass into law. That is to say, the parliamentary stage is now not merely short-circuited, it is omitted altogether. It is an ingenious device for getting the Revised Prayer Book made the law of the realm — for bringing home to port that much-buffeted vessel of the bishops still at sea. But is there the slightest likelihood that Parliament will consent to this abrogation of its statutory powers, one might add — so long as the Establishment remains — its statutory duties? One cannot think so. Professor Ernest Barker, in a letter to *The Times* (February 5, 1936), states the fundamental objection to this proposal from a constitutional point of view:

The effect of this proposal . . . is that a measure which has not gone through the King-in-Parliament, but only through what I may call the King-in-Vacuo, has none the less the force of an Act of Parliament.

He notes the superficial analogy of the proposal with the procedure under the Parliament Act by which the Speaker certifies Money Bills; and proceeds:

But under the Parliament Act the House of Lords has at any rate the opportunity of consenting; and — what is far more important — the House of Commons certainly acts, and the Act is at least an Act of the King in a House of Parliament. A measure with the force of an Act of Parliament, which had never been before either House of Parliament, would be a constitutional novelty, as well as an oxymoron.

He notes further the difficulty there would be in distinguishing between "spiritual" and "secular" measures, since all "spiritual" measures must affect secular interests, and the Commission's own words earlier that the distinction " becomes increasingly unreal . . . there is no clear dividing line in the sphere of practice ". He suggests that the Royal Assent in these circumstances would inevitably mean the assent of the Cabinet; and why should the Cabinet "have powers in spiritual concerns which are denied to Parliament? Could it have such powers, and would it not be responsible to Parliament? If it were, as surely it must be, then Parliament, expelled by an illusory pitchfork, quietly returns." Nor is *The Times* any more convinced than the Professor; its leader on the debate in the Church Assembly (February 8) states: " It is indeed a strange feature of the Report that, having stated the case against disestablishment with admirable clarity, it should then make proposals which must almost certainly be preliminary steps to disestablishment ".

The main feature in the debate in the Church Assembly, according to *The Times*, was the categorical statement of the Archbishop of York — whose speech was greeted with great applause — that "if Parliament, as seems almost certain, should refuse to accept this proposal, then the church should demand disestablishment ". The Bishop of Durham, in a notable speech, thought also " that there was not the remotest possibility of securing the assent of Parliament ". He went further and broached the topic of disestablishment itself:

The risks of disestablishment were greatly exaggerated. It might fairly be assumed that Parliament would be greatly influenced by its own precedents in Ireland, Wales, and India. Parliament was not an inequitable assembly, and due consideration would be given to every legitimate desire and interest of the church. As regards fabrics, the great cathedrals and parish churches were not only houses of God, but the most precious national memorials, and it was reasonable to think that Parliament would hold a watching brief over them for the nation.[1]

It is clear that the submission of these proposals may be likely to accelerate the movement for disestablishment from within the Church. That being so, it is surprising that the Commission should not have realised that disestablishment is the logical conclusion of their Report, though, of course, they were precluded from recommending it. They state their conviction, however, that an agreement among all parties within the

[1] *The Church Times*, February 7, 1936.

The Dilemma of Church and State

Church upon Reservation and a new Office of Holy Communion is a *sine qua non* of their proposals. They realise that it was disagreement within the Church that caused the Revised Prayer Book to be rejected, and they state that their " legislative proposals are made on the assumption that a sufficient measure of agreement has been reached within the church on these two vital questions ". It is a large assumption, and one that is unlikely to be fulfilled. If that is so, what is the alternative? Only disestablishment remains as the way out.

The Commission is conscious of this. It says:

> One method of escape from the entanglements of the whole situation is disestablishment, by which we understand a complete official dissociation of the state from the church. Its attractions are great and obvious, although . . . the practical consequences involved would depend upon the form which disestablishment might take. Disestablishment, it may be urged, could provide a solution of all the difficulties that arise from the conflict of authorities in spiritual matters. It could give the church freedom to determine its own law of worship, and the methods of securing obedience to that law. It could remove the difficulties inherent in any Final Court of Appeal in ecclesiastical cases which owes its character and authority to the civil power. It could set the church completely free to lay down its own rules with regard to the appointment of bishops and in respect to the marriage of its members.[1]

After such a generous and — who can doubt? — true statement of the advantages to be derived from disestablishment by the Church in living its own free spiritual life, it seems difficult to see why the objections to it have such weight with the Commission. Most of the objections appear so trivial: for instance, " no one can be sure ", it says, " that the state would not claim to exercise control over our ancient cathedrals and parish churches, or (to take one obvious case) over Westminster Abbey ". It is clear that the fabric of these buildings would be at least as well off under the Office of Works as under their respective deans and chapters. The State could afford to be generous : to maintain the fabric of these buildings in repair, while allowing their use to the Church for its services —

[1] *Report.* Vol. I, p. 48.

The End of an Epoch

following the model of the French law of 1906 — would be a concession of no small value to the religious body which enjoyed it.

But can it be that there is some other, less "spiritual" reason for rejecting the solution of disestablishment? The Commission says:

> We are bound to recognise that Parliament might insist on some measure, at least, of disendowment as the concomitant of disestablishment. We do not think that there is any necessary connection between disestablishment and disendowment. Nor do we think that the fear of disendowment should be allowed to play any great part in shaping the policy of the church.

These are brave words; but it does not appear incorrect to suppose that this is precisely the consideration that does shape the policy of the Church, that if it were not for the fear of disendowment, the bishops and clergy would opt for disestablishment at no distant date. Such appears to be the view of the Church Association, which roundly contradicts the Commission's complacency on this point:

> Some dream of disestablishment without disendowment. It is only a dream. The English Church will never be disestablished without *drastic* disendowment. The Irish and Welsh churches escaped lightly because they were disestablished simply on the politicians' plea that "the people" willed it. If the Church of England ever be disestablished, it will be because the ecclesiastical rulers have become so dissatisfied with the formularies they have repeatedly of their own free will declared to be agreeable to the Word of God, that they will force even a rupture with Parliament rather than leave church people in peaceable possession of the Prayer Book bequeathed to them by the Protestant Martyrs. Disestablishment would then come as a *punitive* enactment; and an unfaithful church cannot expect satisfactory alimony. Those who imagine that disestablishment means that unfaithful clergymen will be freed from state control and will yet continue to hold the emoluments they possess, are living in a fool's paradise.[1]

There speaks the rude voice of the Church Association. But Sir Thomas Inskip is of the same opinion:

[1] *Report*. Vol. II, pp. 56-7.

The Dilemma of Church and State

If the church feels it (the present position) is a sufficiently great evil to be intolerable, the church is perfectly free, provided it is sufficiently united and insistent, to assert its independence. It is quite true that it cannot carry its property with it if it chooses to take that course.[1]

It is curious the way these Low Churchmen have of insisting upon these low considerations; but there is the dilemma. The dilemma in the relations between Church and State appears to be a dilemma within the Church itself; and it is for the Church to make the first move.

In studying history one becomes used to watching the way in which movements and events prepare the way for their own fruition, while the actors in them remain hardly conscious, if at all, of their trend and direction. One cannot study the history of the Church of England in the past twenty years without observing its steady, if slow, march towards disestablishment. All its main measures and developments in those years have been steps towards it, so that when the time is ripe the way will have been prepared and made easy. The formation of representative institutions for itself in the Church Assembly, the compiling of an electoral roll of its own communicants, the self-confession that the Church is no longer coterminous with the nation;[2] the Enabling Act and its shortening of the legislative process for Church measures, the Revision of the Prayer Book and the fact that the bishops are now administering it in their dioceses as the law of the Church, upon their own authority and without the authority of the State: what are they all but steps towards ultimately freeing the Church from the bondage of the Establishment? The development of what an enthusiastic, if Erastian, college

---

[1] *Ibid.* p. 103.

[2] Cf. H. H. Henson, *Disestablishment*, pp. 42-3: "Out of a total population which exceeded 36,000,000, no more than 3,686,422 persons above the age of eighteen registered themselves in 1927 as members of the National Church, and of these only 2,528,391 were communicants.... How petty a proportion of the English people is included in the membership of the English Church! When due allowance has been made for those communicants who are under age, it may be doubted whether more than one in thirteen of the parliamentary electors is a communicant Anglican." The position since 1927, it need hardly be said, has not improved.

The End of an Epoch

chaplain called the "bureaucracy of Lambeth" means that when the moment of departure comes, there will be no great wrench, for all the instruments will be there for running the Church when the supremacy of Parliament has gone. These events have their logic; and in their movement towards their end, the Archbishops' Commission and its Report, taken with its reception and the effects of it, form yet another significant step on the way.

XII

SOCIALISM AND MR. KEYNES

[1932]

MANY people must have wondered why Mr. Keynes is not a Socialist; and many Socialists, at one time or another, must have had hopes of his conversion. For at almost every point in his economic work where he touches on the same issues as Socialism he suggests similar conclusions. The implications of his general attitude on economic affairs, as well as in pure theory, are closely allied to Socialism; and his ideas on economic policy exercise a greater influence on the Labour Movement than those of probably any other single person. Yet he is no Socialist. Why not? It is a curious problem.

Nor is it the only problem; it deepens into the larger and more important question, Why is it that the most brilliant economic mind of our time, who exerts so wide an influence on economic opinion, has had so little effect upon action? His advice has been persistently neglected by those in authority at every major turning in our policy since the war; and events have proved that he was in the right, and that they were wrong — in some cases, as over our return to the Gold Standard, disastrously wrong. How to account for this failure to be effective?

That Mr. Keynes is aware of it is indicated by his bringing together this year his various pronouncements on economic policy,[1] to stand as a record of his advice and warnings on the economic issues of the post-war years. The record has an extraordinary quality of dramatic excitement in the reading: one watches Mr. Keynes hard at work, while the issues are impending, to persuade bankers, statesmen, industrialists to be reasonable in the interests of their own system, if not in ours; to apply intelligence to the problems that confront them, not

[1] *Essays in Persuasion.* Macmillan, 1932.

Mumbo Jumbo and the rule of thumb. Alas, so much in vain; for, one after another, one sees the decisions made fatally wrong. And then one notices the persuasive spirit of the publicist becoming embittered at so much futility in high places, and the sense of helplessness and defeat growing in the course of the record; until in the end, surveying his commentary on the dreary mistakes of a second-rate decade, he seems almost resigned to being ineffective, and, though aware of the problem why it should have been so, unable to explain it: " Here are collected the croakings of twelve years —" he writes, " the croakings of a Cassandra who could never influence the course of events in time ".

The problem, though it has many facets, is really a political one. It goes down at bottom to a mistaken conception of political activity, one that is far too rationalistic, and lays an exaggerated emphasis on the part played in politics by reason and opinion. It is this that leads Mr. Keynes to attach the importance he does to appealing to the intelligence of those who direct affairs; whereas more important in politics than to hold right views is to place yourself in touch with the group interest that will make your views, when right, effective. Consequently in political tactics he has followed the hopeless line of allying himself with Liberalism, the *laissez-faire* presuppositions of which he has rejected, and of refusing, for reasons which are entirely inadequate when they are not frivolous, to put himself into relation with his effective environment, the Socialist Movement, to which his economic views logically lead him and whose economics is mainly in line with his.

Let us take what I may call the rationalist fallacy underlying his views of politics. He is always bent on appealing to Public Opinion — an attitude all the more quixotic because he never goes into what sort of thing public opinion is, if there is such a thing. " In one way only can we influence these hidden currents (*i.e.* which shape events) — by setting in motion those forces of instruction and imagination which change *opinion*. The assertion of truth, the unveiling of illusion, the dissipation of hate, the enlargement and instruction of men's hearts and minds, must be the means." No wonder he finds himself so

Socialism and Mr. Keynes

ineffective; such an estimation of the forces at work in society, though it does every credit to his character, is almost wholly illusory. He is too much the rationalist. He sees what is reasonable to be done, and then wonders that nobody does it. Of course, they will not; that is not the way things work in society: it may not be to people's *interest* to do it, and that is what counts. There are so many interests pulling them in a different direction; if you want to understand how society works, investigate first those interests: which is what the Liberal mentality never likes to do — as if there is something indecent in the spectacle of economic interest at work.

Besides, such a mistaken estimation of forces is bound to detract from one's effectiveness, and may lead to defeat. In 1929 Mr. Keynes, in placing his policy of expansion to cure unemployment before the Liberal Party, promised them that " Over against us, standing in the path, there is nothing but a few old gentlemen tightly buttoned up in their frock-coats, who only need to be treated with a little friendly disrespect and bowled over like nine-pins ". It reads very funnily; but it was Mr. Keynes who was bowled over like a nine-pin, and the old gentlemen who won — as they will always win, as long as Mr. Keynes sees the conflict in those terms.

It is curious that Mr. Keynes, whose economic views are Socialist in their implication, should in politics place his hopes upon Liberalism. It is true that from 1906, and perhaps before, Liberalism has had a genuine impulse to social legislation, and an advanced Left Wing which was prepared to cooperate with Socialism. But it is impossible for anyone with a historical sense to believe that Liberalism can be persuaded into changing its historic nature. As a party based essentially upon the lower middle class, the shopkeepers and smallholders, it stands, as it always has stood, for the consumer's interest and free trade. Its outlook is *laissez-faire*. Mr. Keynes set himself manfully to transform this outlook after the war to one more in accordance with the changed conditions. Discouraging as it must be to try and get the leopard to change his spots, it is no more hopeful an occupation with a tame tabby. Mr. Keynes had a factitious and temporary success with Mr.

The End of an Epoch

Lloyd George, who, fishing in a bottomless pool for the shrimps of office, was glad of anything in the nature of bait and leaped at the *Liberal Yellow Book* and the *Unemployment Scheme*. These, as publications, are not to be despised: they have all the brilliant inventiveness of Mr. Keynes's mind. The shortcomings were the shortcomings of his political judgment: first, to suppose that you have only to lay before the electorate a brilliant policy and it will respond; second, to imagine that historic Liberalism could transform itself into the exact opposite of all that it has ever stood for. For the purposes of the election, it made the attempt; the results were not encouraging to further experiment. It is clear now that such Liberalism as survives has reverted with Sir Herbert Samuel to *laissez-faire*; that Mr. Lloyd George's leadership has come to an end, and with it Mr. Keynes's hopes of reviving Liberalism on Socialist lines. The most surprising thing is that he should have continued so long to cherish such hopes of the Liberal Party.

In the realm of pure theory Mr. Keynes has been the foremost critic of *laissez-faire*, sapping and undermining the supports upon which, intellectually, Liberalism rests. As early as 1923, in the *Tract on Monetary Reform*, he wrote: " I think that it is not safe or fair to combine the social organisation developed during the nineteenth century (and still retained) with a *laissez-faire* policy towards the value of money ". In his *Economic Consequences of Mr. Churchill*, two years later, he develops the same point in relation to wages policy:

The truth is that we stand midway between two theories of economic society. The one theory maintains that wages should be fixed by reference to what is " fair " and " reasonable " as between classes. The other theory — the theory of the economic Juggernaut — is that wages should be settled by economic pressure, otherwise called " hard facts ", and that our vast machine should crash along, with regard only to its equilibrium as a whole, and without attention to the chance consequences of the journey to individual groups.

After this, it was not unexpected (though at the time it caused a sensation) that he should have declared a final breach with the system of *laissez-faire* principles in his lecture at Oxford: *The End of Laissez-Faire*. In it he invited us

Socialism and Mr. Keynes

to clear from the ground the metaphysical or general principles upon which ... *laissez-faire* has been founded. It is *not* true that individuals possess a prescriptive " natural liberty " in their economic activities. There is no " compact " conferring perpetual rights on those who Have or those who Acquire. The world is *not* so governed from above that private and social interest always coincide. It is *not* a correct deduction from the principles of Economics that enlightened self-interest always operates in the public interest. Nor is it true that self-interest generally *is* enlightened; more often individuals acting separately to promote their own ends are too ignorant or too weak to attain even these. Experience does *not* show that individuals, when they make up a social unit, are always less clear-sighted than when they act separately.

It is an heroic effort to throw off the presuppositions of a lifetime, to think himself free, if such a thing is possible, from the cast of mind which has moulded his own. It is a manifesto against *laissez-faire*; but it is a negative manifesto. He says in it what it is that he does not accept; if you piece out the pattern of principles that he rejects, what is there that remains positively except the general principles upon which Socialism rests?

There is, in any case, a large measure of agreement between him and Socialism with regard to current policy. On all the three major controversies of the post-war period, into which Mr. Keynes " plunged himself without reserve " — the Treaty of Peace and the War Debts, the Policy of Deflation, and the Return to the Gold Standard — he has had more consistent support for his point of view from Socialists than from anybody else. Similarly he has been in sympathy with their case with regard to the social services; he has made their preservation and development a main consideration in his resistance to deflation and its consequent creation of unemployment. He has pointed out the unfairness of its incidence upon different classes in the community; he has shown how the coal miners have been peculiarly the victims of our monetary policy. Similarly, at the time of the May Report, and again with the National Government's Economy Bill of 1931, he protested with indignation at the injustice and unwisdom of placing yet worse burdens on the most depressed classes while still further contracting purchasing power and productive enterprise. His part in all these conflicts reveals a remarkable sympathy for the working class, and a rare understanding of how the economic impact of capitalist society weighs most heavily on them in the end.

Again, with his conception of the public concern as a model for large-scale enterprise in the transition to Socialism, he is in line with current Socialist opinion. Mr. Keynes popularised the conception in the *Liberal Yellow Book*; but it was Mr. Herbert Morrison who gave these ideas their practical embodiment in the London Passenger Transport Bill. He arrived at this conception by the same path as Socialist critics of capitalism; like them, he observed " the tendency of big enterprise to socialise itself. . . . We see here, I think, a natural line of evolution. The battle of socialism against unlimited private profit is being won in detail hour by hour." He agrees that the trend has not gone far enough and must be carried farther: " It is true that many big undertakings, particularly public utility enterprises and other business requiring a large fixed capital, still need to be semi-socialised ". And he suggests

that progress lies in the growth and the recognition of semi-autonomous bodies within the State — bodies whose criterion of action within their own field is solely the public good as they understand it, and from whose deliberations motives of private advantage are excluded — bodies which in the ordinary course of affairs are mainly autonomous within their prescribed limitations, but are subject in the last resort to the sovereignty of the democracy expressed through Parliament.

His scheme begins to look oddly like Guild Socialism: " I propose a return, it may be said, towards medieval conceptions of separate autonomies ". Certainly it is open to the chief objection against Guild Socialism — its inadequate provision for the coordinating functions of the State. But even more strongly; for at this point there is a lacuna in Mr. Keynes's ideas, and he provides no means of integrating these various discrete corporations and bringing them effectively into relation with the general political structure. Again, it is his conception of politics that is lacking, a failure to draw out the necessary political conclusions from the right economic premises. It may be said that he looks to a measure of monetary control for this; but it is not monetary control that would effect this particular purpose: it needs the integrating power of the State itself, on a very comprehensive conception of the State's powers. In short, it requires a Socialist State.

No one has pursued the campaign for monetary control in itself with more constancy and force; the idea that this country could subsist on a managed currency at all may almost be said to be Mr. Keynes's creation. Very early on, before our return to the Gold Standard, he was arguing for a system of monetary control which should be manipulated in the interest of productive enterprise and employment, instead of that of the rentier class and of our foreign lending. Such " a deliberate control of the currency and of credit by a central institution " he regarded as the best means to cure many of the greatest economic evils of our time — the great inequalities of wealth, unemployment, the slowing down of production owing to the failure to expand either profits or demand. And there is the further corollary of this control, in the field of savings and investment:

I believe that some coordinated act of intelligent judgment is required as to the scale on which it is desirable that the community as a whole should save, the scale on which these savings should go abroad in the form of foreign investments, and whether the present organisation of the investment market distributes savings along the most nationally productive channels. I do not think that these matters should be left entirely to the chances of private judgment and private profits, as they are at present.

In the later *Treatise on Money* he elaborates the view that the lack of coordination between savings and investment is a main cause of trade depression; and he urges that the decisions as to the proportions of the flow of future output should be made by the same people who decide how much is to be saved. Without prejudice to the theoretical question, what is this but what the Russians in fact do?

With this he definitely crosses the frontier between capitalism and Socialism. Up to this point his views, though in line with Socialist thought, might reasonably be regarded as not going beyond the bounds of the new Liberalism. But with this control over foreign investment he is laying his hands on the central citadel of capitalism. For some reason he regards this advance as merely an " improvement in the technique of modern capitalism ", and not at all incompatible with it. I suspect it to be due to his not defining capitalism clearly enough, for he says that its essential characteristic is " the dependence upon an intense appeal to the money-making and money-loving instincts of individuals as the main motive force of the economic machine ". Surely this is not an adequate differentia ? For one can imagine a régime based on the appeal to money-making which was not capitalistic for all that. Is it not rather the fact that this society based on money-making is manipulated by individuals who are irresponsible to it that makes the difference ? So that one might, theoretically, retain the appeal to money-love, but operating in the conditions of an egalitarian society ; and it would not be capitalism. The difference is constituted by the fact that power, in a capitalist society, is the result of economic inequality, and is held irresponsibly — that is to say, on terms laid down by itself, and not by society. However, it is a far more serious consideration that divides him from capitalism on the issue of savings and investment. It is a struggle of power ; and for all his seductions, capitalists are not likely to give up the key to the control of their system.

But he has some uncomfortable words to say to them concerning this indispensable part of the machine :

> The system [*i.e.* of foreign investment] is fragile ; and it has only survived because its burden on the paying countries has not so far been oppressive, because this burden is represented by real assets and is bound up with the property system generally, and because the sums already lent are not unduly large in relation to those which it is still hoped to borrow. *Bankers are used to this system, and believe it to be a necessary part of the permanent order of society.* . . . The practice of foreign investment is a very modern contrivance, a very unstable one, and only suited to peculiar circumstances.

But it lies at the root of capitalism in its most developed form ;

Socialism and Mr. Keynes

economic imperialism could not survive, in the form we know it, without finding foreign markets in which to invest its surpluses and from which to draw its raw materials. What, then, does Mr. Keynes think of the future of capitalism in this country without the stress on foreign investment? He might answer that he looks to the time when most of these surpluses will be devoted to raising the standard of living inside the country and to re-equipping industry; in that respect his aim is the same as the Bolsheviks'. But does he think that can come about, without the State having a great deal more control over private capitalism than there is in being or than capitalism allows for? True, he is in favour of some such instrument as a Board of National Investment to control the allotment of surpluses as between home and abroad, industry and industry. But so far there is no disposition whatever on the part of capitalists to admit the insufficiency of their own system or to agree to any control of investment by a State Board such as Mr. Keynes has suggested.

There are other reasons for regarding capitalism as unstable, without some such control as would transform its nature — reasons connected with the general mechanism of money and credit, which no one has done more to elucidate than Mr. Keynes. In the *Tract on Monetary Reform* he first investigated this element of instability due to the dependence of the system upon a monetary standard that is liable to great fluctuations in value; he estimated the social consequences of these changes and showed that it is precisely because capitalism leaves saving and investment to the individual, without imposing a social control, that the instability of money values is liable to be all the more disastrous to the whole machine. " The Individualistic Capitalism of today, precisely because it entrusts saving to the individual investor and production to the individual employer, *presumes* a stable measuring-rod of value, and cannot be efficient — perhaps cannot survive — without one." The post-war period has revealed the insecurity of the foundations upon which capitalism was built; and even the most convinced believers in the eternity of the system must be considerably shaken by now. The effect has been electric in every direction; to take only the most obvious of its consequences: the throwing out of gear of the relations between creditor and debtor countries; the similar maladjustment between creditor and debtor interests *within* countries; the disturbance to the expectation of profit which has had such a destructive effect upon productive enterprise. Any one of them is enough to deal a severe blow at the stability of the system. In order that enterprise may be given a normal expectation of return for its work, Mr. Keynes would provide for the redistribution of wealth when necessary, as he would have welcomed a capital levy after the War of 1914-18, to reduce the claims of the rentier interest upon production. On the question how far capitalism can survive the instability of its monetary standard he says:

> Modern Capitalism is faced with the choice between finding some way to increase money values towards their former figure, or seeing widespread insolvencies and defaults and the collapse of a large part of the financial structure; — after which, we should all start again, not nearly so much poorer as we should expect, and much more cheerful perhaps, but having suffered a period of waste and disturbance and social injustice, and a general rearrangement of private fortunes and ownership of wealth.

This passage might be described as the Prelude to Socialism. Nor does Mr. Keynes differ from Socialists in his conception of the transition from capitalism to what can only be called socialism: " the transition from economic anarchy to a régime which deliberately aims at controlling and directing economic forces in the interests of social justice and social stability ".

In what, then, consists the difference between Mr. Keynes and most Socialists? It is very hard to find any fundamental divergence of principle in their aims, or in the way they envisage the economic problems of society. Perhaps it may be said that they diverge in their estimation of the importance of the entrepreneur and the treatment he would receive in a Socialist State. But this is doubtful. Mr. Keynes sets great store by the active spirit of business enterprise. Doubtless it is very precious; and under the existing system it operates best through the encouragement offered by monetary rewards. But

Socialism and Mr. Keynes

that does not mean that without these it would not operate at all. Power is what it wants most; and if given power it should have no difficulty in working within the social conditions prescribed by the system. If it wants power, it should be prepared to make such sacrifice of the mere accessories of power under the existing order as would be necessary under the conditions of the new. Surely it would be willing to give up its conventional standard of living, to forgo the comforts of its present social *milieu* if necessary? The point is whether, given the chance of reorganising the heavy industries of the country, or running the Bank of England or the traffic of London, you would not be prepared to give up your house in Park Lane in order to qualify under Socialism. No one with the real desire for executive power would hesitate for a moment — any more than a great manager like Krassin did on the outbreak of the Russian Revolution. Still this remains a very important issue for Socialism: upon what terms the administrative and managing class in this country will come over to it. The system of differential rewards no longer functions properly on the old basis. Under joint-stock enterprise, particularly in its later developments, responsibility can no longer be brought immediately home to the individual business man, as capitalism assumes. If he is inefficient, he goes on being inefficient; he probably does not become a bankrupt. So that we have under the present system — it is a notorious cause of inefficiency — what the *Liberal Yellow Book* called " the pocket-boroughs of industry ". It is all very well to point them out to a laggard public; but what hope is there of dealing with them unless by transforming the system?

Mr. Keynes has a soft spot in his heart for the entrepreneur; and this affection leads him at times to over-step his own logic and allot him a larger part in the economic system than his premises allow. For example: " The intensity of production is largely governed in existing conditions by the anticipated real profit of the entrepreneur. Yet this criterion is the right one for the community as a whole only when the delicate adjustment of interests is not upset by fluctuations in standard of value." Is this a good enough criterion, indeed?

The End of an Epoch

Of course not; there are other conditions to be fulfilled besides stability in the standard of value before it can be admitted that the profit of the entrepreneur is the criterion of the community's welfare. We can all agree with his recognition of the constructive capacity of the entrepreneur class. In his criticism of Mr. H. G. Wells's *Clissold* he underlines Mr. Wells's intuition that a constructive revolution cannot possibly be carried through by the sentimentalists and pseudo-intellectuals of the Labour Movement, who have " feelings in the place of ideas ". In Mr. Wells's view, at least in *Clissold*, " the creative intellect of mankind is not to be found in these quarters but among the scientists and the great modern business men. Unless we can harness to the job this type of mind and character and temperament, it can never be put through — for it is a task of immense practical complexity and intellectual difficulty." But he sees more clearly than Mr. Wells that " they lack altogether the kind of motive, the possession of which, if they had it, could be expressed by saying that they had a creed. They have no creed, these potential open conspirators, no creed whatever." And even on a more restricted ground, whether the aims of the business man should prevail in the organisation of society, he is with us : " We doubt whether the business man is leading us to a destination far better than our present place. Regarded as a means he is tolerable; regarded as an end he is not so satisfactory." In the result, his criticism of capitalist society becomes this, that though he regards it as more efficient technically than any other alternative system, he regards it as inadequate on social grounds.

For my part, I think that Capitalism, wisely managed, can probably be made more efficient than any alternative system yet in sight, but that in itself it is in many ways extremely objectionable. Our problem is to work out a social organisation which shall be as efficient as possible without offending our notions of a satisfactory way of life.

One wonders whether he is so convinced now, as when he wrote that, that capitalism can be made so much more efficient and yet remain capitalist? As for the problem, no Socialist would take exception to his statement of it.

Socialism and Mr. Keynes

Why, then, does he hold aloof in this curious, and apparently inexplicable, isolation? In my view, it is due to a defect of political vision — a defect which, I hasten to add, is not irreparable. When he writes on politics he displays little of the characteristic excellences of his technical work. He is liable even to make a howler — as when he identifies public opinion with Rousseau's General Will. He would hardly have committed this crime of mistaken identity if he had been brought up at Oxford instead of at Cambridge. Rousseau's notion of the General Will is a much more profound concept; it is something much more allied to the ideal of the common interest of society. Whereas public opinion means — well, what? Mr. Keynes has the idea that the function of the publicist is to be concerned with what public opinion should be, to form it rather than to take his cue from it. But ought not the first business of a writer on public affairs be to know what public opinion *is*, and how it is made up? He would find it a reflection of group interests, lagging somewhat behind them in its expression of their needs. Sometimes so much so that it may even cease to serve the ultimate interest of the group it is intended for. However, it remains, whether enlightened or not — more often not — a psychological state moulded in accordance with the class organisation of society. It is merely an example of the rationalist fallacy to suppose that it moves by its own volition.

This excessive rationalism of outlook, to which Liberals are peculiarly susceptible — Tories have no illusions in the matter — has made him fail to understand the relations that subsist between party and class. His objection to the Labour Party is that it is a class party. The Liberal Party is not a class party: it is " disinterested as between classes ". One has heard it so often that it might appear to be a platitude — if it were not an obvious falsehood. All parties are class parties; the differences between them are the differences between the classes of which they are constituted. It is absurd to complain of the Labour Party that it appeals to the class feeling of the working class. Would indeed that it did a little more forcefully! Is it necessarily a bad thing to appeal to class feeling? Would

The End of an Epoch

the régime of feudal privilege have been abolished in France by the Revolution if it had not appealed to the class hatred of the bourgeois for the aristocracy? That may be appealed to, it seems: at least, it is allowed by all Liberal opinion; but to the legitimate feeling of the working class, not. This frame of mind goes along with the defence for themselves — everybody is always ready with a defence for himself — that *they* care particularly for Truth. They want to find out what the truth is — they want to be free of the connections and the class associations which prevent other people from the pure service of the truth. As if such a state of theoretical nudity were possible, however desirable it may be! We also want to get at truth; though we have not so simple and naïve a view of its nature. We know how it is always the people who are in search of pure truth, who neglect the bends that interest — both of the group and of the self — exerts upon the mind, that are always the most blind to the way in which class associations affect their own outlook. The only remedy is to bring all such economic and psychological bends into the foreground and make them explicitly a part of your thought.

Mr. Keynes's point of view is far too realistic for this to be specially applicable to him; it rather marks him off from the common ruck of Liberals. When it comes to the point, he can give expression to a realistic view that is refreshing, if crude.

Ought I then to join the Labour Party? Superficially that is more attractive. But looked at closer, there are great difficulties. To begin with, it is a class party, and the class is not my class. If I am going to pursue sectional interests at all, I shall pursue my own. When it comes to the class-struggle as such, my local and personal patriotisms, like those of everyone else, except certain unpleasant zealous ones, are attached to my own surroundings. I can be influenced by what seems to me Justice and good sense; but the *Class* war will find me on the side of the educated *bourgeoisie*.

Now this is altogether too restricted a view; its danger is that, just for lack of making one further step, it may land its holder in pure cynicism. Which would be a pity; for his great gifts should be at the service of the community. But in these terms of interest and class it is possible to understand each other.

Socialism and Mr. Keynes

Let us try now to close the gap. How if the interest of the working class were more closely identified with that of the whole than any other sectional interest? How if it were the only possible basis for that drawing together of the closest bonds between peoples which Mr. Keynes has worked so devotedly for? So devotedly, and so much in vain; for it looks as if it is not possible to bring about any real cooperation through the governing classes of the world, to whom Mr. Keynes has exclusively looked in the past. Can it be that the only lasting foundation is to be found in the working classes of the world? *Workers of the world, Unite?* If so, if it can only be brought about on this basis, then it needs the concentration of all the ability we have to the task. But the most brilliant mind in England remains sulking in his tent.

It is only a figure of speech and may be thought an unjust one. For, indeed, Mr. Keynes can claim that he has struggled manfully for right and justice, and more consistently than most, on all the main economic issues since the war. More, he has been right about them, and in the teeth of prevailing opinion. But that is precisely the tragedy of the situation. It is a melancholy satisfaction for a man to know that he has been right all the time, and that nobody in power took any notice of what he said. Why should this be? He has been unfortunate in that he was a Liberal in time of the decay of Liberalism. But it goes deeper than that; nor is he without fault in the matter himself. He has not cared to bring about a right relation to the one movement that could make him effective. It is not only exact thinking, good leadership, and a programme that are necessary. To suppose that they are enough is the kind of mistake a rationalist, an intellectual, would make. In making his appeal to Liberalism he offered all these: " We must take risks of unpopularity and derision. *Then* our meetings will draw crowds and our body be infused with strength." They took these risks: the crowds did not come; for crowds do not respond to programmes. They respond to *interests* that they have in common; it is the work of the real leader to clarify these and to direct them to the right social ends. As for ideas and programmes, the movement is more important than either; what it is will determine what it will do. And if the intellectual is really to be of use, he will use his intellect, not to make superficial points, like those that are so easy to make against the Labour Movement — and Mr. Keynes makes them all — but to understand what it implies and whether by working through the working-class movement the changes he desires, along with us, may come about. If he begins to consider this, he may realise that they will not come about on any other basis.

It is difficult to see that there is any rational reason for his refusal to cross the frontier in politics, when he has already done so very amply in economics. It must be due to some emotional inhibition. The emotional reaction is impossible to mistake in what he has written with regard to Communism. " How can I accept a doctrine ", he says, " which sets up as its bible, above and beyond criticism, an obsolete economic textbook which I know to be not only scientifically erroneous but without interest or application for the modern world?" Can it be that he has been defeated in the attempt to read Marx, or that he has never tried? I have long suspected the one or the other. It is certainly very easy to be defeated by *Das Kapital*, though it is no more out of date than, say, Mill or Ricardo. But it is a purely emotional judgment to say that it is " without interest or application for the modern world ": would he be so certain of it now as he was then? Or again: " How can I adopt a creed which, preferring the mud to the fish, exalts the boorish proletariat above the bourgeois and the intelligentsia who, with whatever faults, are the quality in life and surely carry the seeds of all human advancement?" That "surely" strikes a note of inner uncertainty. Of course, the proletariat are sadly lacking in all that they should be, viewed from the region of Bloomsbury. But for all that, theirs may be the creative role in modern society; when his own bourgeois are losing their constructive power in politics, and nobody has demonstrated their economic failures better than he has himself. When Christianity first arose, to be a world movement, it sprang, not from the Pharisees, nor even from the Sadducees, but from a handful of doubtless very unsatisfactory fishermen.

Socialism and Mr. Keynes

True, it was given the organisation it needed by a thorough bourgeois, St. Paul; but it is a role that is always open to be filled by any bourgeois that volunteers, though Lenin has staked out rather an impressive claim to it.

And one wonders how much of the ill-balanced cocksureness of 1925, as to the significance and achievement of Communism, has survived with the years. " On the economic side I cannot perceive that Russian Communism has made any contribution to our economic problems of intellectual interest or scientific value." Would he be so certain of it now, or of this ? —

I do not think that it contains, or is likely to contain, any piece of useful economic technique which we could not apply, if we chose, with equal or greater success in a society which contained all the marks, I will not say of nineteenth-century individualistic Capitalism, but of British bourgeois ideals.

For it may be that these same bourgeois ideals forbid any significant change in the nature of the system, and that for that you will have to look to the working-class movement for leverage.

It is not dissimilar, as we have seen, with his attitude to Socialism in England. He says of it: " Socialism offers no middle course [*i.e.* between the old *laissez-faire* and the new régime of control], because it also is sprung from the presuppositions of the Era of Abundance, just as much as *laissez-faire* individualism and the free play of economic forces ". A pedantic point, which can be replied to equally pedantically by pointing out that it sprang from precisely those classes that enjoyed least abundance. Of the inner constitution of the Labour Movement he has, like most Liberals, a fantastic notion ; the Trade Unionists with the Communists cheek by jowl play nefarious roles.

The Labour Party contains three elements. There are the trade unionists, once the oppressed, now the tyrants, whose selfish and sectional pretensions need to be bravely opposed. There are the advocates of the methods of violence and sudden change, by an abuse of language called Communists, who are committed by their creed to produce evil that good may come, and since they dare not concoct disaster openly, are forced to play with plot and subterfuge.

There are the Socialists, who believe that the economic foundations of modern society are evil, yet might be good.

He goes on to add: " The company and conversation of this third element, whom I have called Socialists, many Liberals today would not find uncongenial ". He conceives that there is much sympathy between them, and a similar tendency of ideas; and that they will draw nearer together in constructive work as time goes on. In the end, the only difference he arrives at between Socialism and his own version of Liberalism is " a certain coolness of temper, such as Lord Oxford has, and seems to me peculiarly Liberal in flavour ". One may easily over-estimate the importance of coolness of temper in politics; and anyhow, Lord Oxford is unquestionably dead.

Mr. Keynes's economic ideas, I have shown, imply Socialism. Then why does he not draw the necessary conclusions from his economics in the political sphere? It is due, again, to the exaggerated rationalism of his outlook on politics; he fails to go on from the economic principles he has laid bare to the necessary institutions that embody them in society. It is that typical bourgeois phenomenon, the victory of the technician over the power to conceive society as a comprehensive whole. It is the same rationalist fallacy that leads him to choose his political home on the principle of repulsion, rather than the principle of attraction, and " to go to those whom he dislikes least "; the same fallacy leads him to misconceive the necessary role of Trade Unions in politics and to dislike their association; it leads him to suppose that the unthinking mass of the modern electorate may be won with bright ideas, and their political allegiances changed by programmes; it led him to make his famous invitation over the wireless to the public to go out and spend their cash, the ridiculousness of which ordinary common sense at once saw through; for what was the point of going out and spending all you had while the claims of the debt holder and the rentier and the income-tax collector remained what they were? Yet it is very revealing of the man that his essential economic ideas lead him to a position that he rejects on emotional grounds, while at the same time, when viewing

Socialism and Mr. Keynes

the mass of the people in politics, whose behaviour is instinctive, barely conscious, and anything but rational, he insists on applying a rationalist technique, fit only for a don's domestic affairs. How provoking the most intelligent of mortals can be! There he remains, betwixt and between — an inescapable influence upon the thought of his time, and with no discernible effect upon action.[1] Yet he will not take the last and obvious step to it. Like Newman at Littlemore, he prolongs the agonies unconscionably, while those who watch outside see more clearly the right and inevitable step. And yet, even now, it is possible he may never make the journey from Littlemore to Edgbaston.

On the other hand, there are some signs he may. He reveals again and again his perception that what society needs is a creed. He seems to understand this need perhaps better than Mr. Wells does; at least, it is the ground on which he criticises *The World of William Clissold*. But when he comes to look into the future himself, the *Economic Possibilities for our Grandchildren*, with all his optimism as to economic progress, he has nothing to say, or but little, as to our spiritual need. We find he is only saying that in the future we shall have more and more material comforts. But if this is all that the Communists meant by their Revolution it would not have been worth the sacrifice. They intended not only the increase of material comfort, but the creation of a Just Society. He combines, then, this pathetic desire for a creed with an inability to believe in one. For such a person there is only an intellectual persuasion that remains. Having seen its necessity, he may be open to persuasion as to the source from which alone it can come; and we may expect a reasonable act of imagination towards a system, not of his own fashioning nor of his own environment, but that is nevertheless inevitable.

---

[1] It took the outbreak of another war to bring the most brilliant economic brain in the country into the field of action and give him full scope — with what inestimable consequences! Without him we might never have borne the financial burden of the war. But why could it not have come about years before?

XIII

THE RISE OF LIBERALISM

[1936]

IT was observable that H. A. L. Fisher's *History of Europe*, that fine achievement of a distinguished Liberal of our generation, ended upon a sceptical, disillusioned note. Indeed, he drew attention to the temporary, tentative character of the Liberal triumph in the nineteenth century, by entitling his last volume " The Liberal Experiment ". In so doing, he explained, he was

using the adjective Liberal in no narrow party sense, but as denoting the system of civil, political, and religious freedom now firmly established in Britain and the Dominions as well as among the French, the Dutch, the Scandinavian and American peoples. And if I speak of Liberty in this wide sense as experimental, it is not because I wish to disparage Freedom (for I would as soon disparage Virtue herself), but merely to indicate that, after gaining ground through the nineteenth century, the tides of liberty have now suddenly receded over wide tracts of Europe.

It is to be noted that among the various freedoms specified, " civil, political, and religious ", there is one that is absent — economic freedom. Can it be that the omission was of set purpose? For when one considers all that is implied in that issue for our society, can one be certain that even the political and social system in those countries is so " firmly " established? It may be that it is; but there can be no doubt of the resounding defeat of that system outside them — and not, it may be added, in Europe alone.

Now that we see plainly the significance of the defeat — for, whether it is Fascism or Communism or some form of Socialism that will emerge from the chaos of contemporary Europe, the one thing clear is that it is Liberalism that has

The Rise of Liberalism

been defeated — we may with profit go into its origins, trace its rise, analyse the conditions necessary to its growth and achievement, attempt to define its character. The moment has been seized by Professor Laski, whose books have more and more a topical reference — it is the source of their verve and popular appeal, no less than of their unsatisfactoriness. Nevertheless, considerable if discursive reading has gone to the making of this new volume;[1] and it is stimulating and suggestive in its point of view, though one could have wished for more spaciousness and repose for reflection. Mr. Laski is conscious of this objection, for he says in his Preface that his book is " essentially an essay " in which " it is impossible to do more than sketch the main outline of the theme "; that, " for adequacy, a much more detailed analysis would be required ". He calls the book an " Essay in Interpretation ". It is an essay, but hardly in interpretation; it is rather an historical essay, providing, he says, the background to its predecessor, *The State in Theory and Practice*.

He does not define what he means by liberalism, but that, no doubt, is a consequence of his historical treatment of the subject. Instead he builds up a conception of the movement of society towards liberal institutions, tracing a concurrent development in the minds of the chief political thinkers from the Reformation to the French Revolution, from Machiavelli and Bodin to Voltaire and Rousseau. It is a well-worn track; and sometimes one wonders whether this is not a history of European political thought rather than a history of European liberalism. But the conception Mr. Laski has of his subject explains this treatment. He regards liberalism as the doctrine of the middle class in its rise to power, the expression of its needs and aspirations on the way — in a word, the rationalisation of its interests. In fact, at the end of his book he declares roundly that as a doctrine it was, effectively, a by-product of the effort of the middle class to win its place in the sun. As it achieved its emancipation it forgot not less completely than its predecessors that the claims of social justice were not exhausted by its victory.

[1] *The Rise of European Liberalism*, by H. J. Laski.

That opens up an important issue; for, while from the point of view of historical struggles and social change a body of doctrine can be conveniently regarded as a by-product, from the point of view of theory and of the values of human experience it may have an importance over and above the historical conditions that brought it into being. Without going so far as to regard it as an end in itself, it is obviously something more than a by-product. Mr. Laski's scepticism is obviously due to the influence of Marxism; but even a Marxist may admit that a doctrine, once it has come into being and been established in certain conditions, may transcend those conditions and exert its influence beyond its original field. That is to say, that over and above the historical factors giving rise to liberalism there may be something in the content of the doctrine, in its message, in its values, that transcends the original conditions and may have an influence in turn upon a new phase of society. In addition to the historical factors giving rise to the doctrine there is the question of its content.

Mr. Laski is concerned with the first aspect of the question, not with the second; indeed, so complete is his historical scepticism that it may be doubted whether the second has any validity or importance in his view. It certainly simplifies the treatment of the subject, but at the cost of what is most fascinating in it intellectually. Nevertheless, in this more limited field, his historical approach is in keeping with the more obvious trends of recent scholarship. It may be said that he lays a disproportionate emphasis upon the earlier period, dealing in detail with the sixteenth and seventeenth centuries, when barely a trait of what became subsequently recognisable as liberalism had as yet emerged; so that the book seems to be dealing with the pre-conditions of liberalism rather than with its actual rise. But let Mr. Laski state his view in his own words:

What produced liberalism was the emergence of a new economic society at the end of the middle ages. As a doctrine, it was shaped by the needs of that new society; and, like all social philosophies, it could not transcend the medium in which it was born. Like all social philosophies, therefore, it contained in its birth the conditions

The Rise of Liberalism

of its own destruction. In its living principle, it was the idea by which the new middle class rose to a position of political dominance. Its instrument was the discovery of what may be called the contractual state. To make that state, it sought to limit political intervention to the narrowest area compatible with the maintenance of public order. It never understood, or was never able fully to admit, that freedom of contract is never genuinely free until the parties thereto have an equal bargaining power. . . . The idea of liberalism, in short, is historically connected, in an inescapable way, with the ownership of property.

There is much that is common form in that today; but is it really necessary, in order to lay bare the origins of liberalism, to go back to the end of the Middle Ages? Or, if it is necessary, why stop there? Some people might think that there was a good deal more liberalism of a sort in the Middle Ages than there was in the sixteenth century, the age of the new despotism, of the union of secular and spiritual sovereignty over half Europe, of thoroughgoing economic regulation in the interests of the nation State: the century of Machiavelli at one end and of Bacon at the other — no liberals, surely? Indeed, Mr. Laski is reduced to admitting this when he confronts the evident facts of sixteenth-century economic nationalism, *étatisme* and mercantilism. He says half-heartedly: " Individual economic good is still set in the context of a community-good of which the State is the appointed guardian. Men are still too accustomed to intervention of authority in economic life to doubt its general validity." More candidly: " We can see, in men like Bacon, at the end of the age, that a strong State rather than a free individual, that *étatisme* rather than liberalism, is still the dominating conception ". " Still " ! — it remained so up to the end of the eighteenth century, as he later admits, giving his whole argument away:

It is customary to call the whole period between the Reformation and the French Revolution the age of mercantilism; and it is certainly true that until the latter part of the eighteenth century there was no wide appreciation of liberalism in the economic field.

So much for liberalism in the sixteenth century. There may be some point in going back to the Reformation, since in history everything depends upon everything else. No doubt in driving a car one presupposes the laws of mechanics, and even the universe in which the car has a place; but it is more usual for practical purposes to take them for granted and concentrate on the question in hand, the actual driving as an end in itself. Perhaps there is this to be said from the point of view of liberalism for going back to the Reformation, that it did make for an appeal away from a universal authority in spiritual matters to the individual conscience. The process was by no means so sudden or so far-reaching as the text-books would give one to suppose. But the substitution of the authority of the Bible among Protestants for that of the Church — a book which was liable to as many interpretations as it had readers — had the effect of producing numerous sects, with whose history, as all would agree, liberalism has been closely bound up. Perhaps it is for this reason also that liberalism has been a phenomenon of Protestant countries, not of Catholic. In Catholic Europe, in Spain, in Italy, Austria, Germany, even in France, it has always had something of an exotic air and enjoyed only the feeble health of the foreign transplantation. Equally significant, where it has sprung up in those countries it has usually been in association with anti-clericalism.

There is more point, therefore, in going back to the Puritan Revolution of the seventeenth century for the origins of liberalism. It is then that we begin to see certain characteristics emerging from the depths of a troubled society, out of civil war and revolution, that in association take the shape we later, and in a more peaceful age, recognise as liberalism. Mr. Laski's treatment of this period, of which he has made special study, is better; and he is quite right to point out that, however peaceable a shape liberalism took on later, it had its birth in revolution. " The liberal need ", he says, " is a doctrine woven from the texture of bourgeois need. It is the logic of the conditions they require for their ascent. The pattern of the creed is set by their necessities." Because the monarchy and the landed nobility were in possession, and imposed their appropriate religious outlook with its emphasis on obedience and divine right, the rising bourgeoisie, the

The Rise of Liberalism

merchants of the towns with their allies among the lesser gentry, were driven to attack the monopoly of the Church. Because these rising classes were out for power, they concentrated their effort upon Parliament and the law, their appropriate political instruments; they made constitutionalism their ideal as against a divine-right monarchy, because it suited their interests. Because the old system of regulation of wages and prices, and the monarchy's customary rights with regard to monopolies and the regulation of trade interfered with their interests, they developed a theory of non-intervention of the State in trade. By the end of the seventeenth century Charles Davenant was writing:

> Trade is in its nature free, finds its own channel, and best directeth its own course; and all laws to give it rules and directions, and to limit and circumscribe it, may serve the particular ends of private men, but are seldom advantageous to the public.

How this points forward to Adam Smith, to Ricardo and Mill and Marshall — the great hierarchy of liberal economists; and how rare it is for even the most penetrating of human minds to realise that their generalisations depend upon particular circumstances, that they universalise what is their own particular interest.

The struggle out of which liberalism was born came to a head, then, in the seventeenth century; and it is not without reason that contemporary Liberals should make much of their ancestors, that Mr. Isaac Foot, like Lord Morley before him, should hark back to Cromwell, for all his sending Parliament packing about its business. The confessional affiliation is more important than constitutional forms; and the cynic might say, now that political Liberalism is practically restricted to Nonconformity, that it would probably have preferred, as the human manner is, the rule of its own creed to the constitutional niceties to which it attached so much importance when fighting for its hand. But that century of struggle ended with a compromise; the new bourgeoisie did not wholly win; it came to terms with the landed oligarchy; together they brought in a new dynasty upon terms and ruled England for two centuries and more. There were many things that went overboard to make that compromise upon which modern England rests: the very real spirit of egalitarianism that came to birth in the armies of the Parliament, the social radicalism of the levellers.

The compromise received, at the very moment of its practical victory, its classical expression in doctrine with the works of Locke. In politics, the idea of a contract between Government and people, with an ultimate right, carefully hedged about and restricted, to overthrow the Government without, of course, dissolving civil society: what is that but a rationalisation of the circumstances of 1688, a defence of the Whigs' political revolution without involving the awkward consequences of social revolution? In economics and social theory, the safeguarding of the rights of property, the derivation of property and capital from individual and personal labour: in religion and education, a strictly constitutional God not interfering with the reasonable concerns of man, a limited tolerance for the Protestant sects: in philosophy, a common-sense empiricism excellently adapted for evading the profounder difficulties into which it did not wish to probe. What other, in fact, than Liberalism fully fledged and tried, ready for its long and creative period of rule in this country and to exert a widespread and sustained influence — no English doctrine has exerted a greater — upon thought and (some) practice abroad? No wonder Mr. Keynes, himself at the end of that long and noble tradition of English thinkers, calls Locke's *Essay Concerning Human Understanding* " the first modern English Book ". It might with equal justice, and more illuminatingly, be regarded as the first classic of the English bourgeoisie.

The record of Liberalism has been a very fruitful and creative one; and, critical as Mr. Laski is of its restricted assumptions, he pays tribute to its historic achievements:

I do not mean [he says] that the triumph of liberalism did not represent a real and profound progress. The productive relations it made possible immensely improved the general standard of material conditions. The advance of science was only achieved through the mental climate it created. All in all, the advent of the middle class to power was one of the most beneficent revolutions in history. . . . No one can move from the fifteenth to the sixteenth, still more to the seventeenth, century without the sense of wider and more creative horizons, the recognition that there is a greater regard for the inherent worth of human personality, a sensitiveness to the infliction of unnecessary pain, a zeal for truth for its own sake, a willingness to experiment in its service, which are all parts of a social heritage which would have been infinitely poorer without them. These were the gains involved in the triumph of the liberal creed.

But a shadow lay upon that triumph, as, indeed, what human endeavour has not the shadow of time and changing circumstance upon it? It is evident to the careful observer even in the sanguine days of the formation of liberal doctrine. As Mr. Laski says, it erected its imposing structure of constitutionalism, *laissez-faire*, and natural rights upon a basis of economic inequality. "Having made inequality an implicit article of its faith, it then invites to freedom those who are denied the means of reaching it." In time the shadow grew larger; beyond the claims of the *Tiers État* to a share in political power loomed the social claims of a Fourth Estate. Its own battle hardly won, the middle class in the midst of the struggle over the French Revolution could not afford to meet the challenge on the Left with concessions. "A doctrine that started as a method of emancipating the middle class changed, after 1789, into a method of disciplining the working class."

Liberalism did not renew its forward march after the experience of that revolutionary epoch until the eighteen-thirties. (Is it not a mistake on Mr. Laski's part to date its renewal of strength in France to 1815? The July Revolution of 1830 is a more convincing sign, and, very significantly, our own Reform Bill followed only two years later.) But the more liberalism in the nineteenth century pressed for the political enfranchisement of the masses, the more it opened the way to their social and economic claims. The restricted connotation of liberal doctrine could not contain them; hence, in Mr. Laski's view, "the declining authority of liberal doctrine in our epoch. It was so preoccupied with the political forms it had created that it failed adequately to take account of their dependence upon the economic foundation they expressed."

The breakdown in doctrine is not difficult to trace. Take the doctrine of non-intervention: in the view of Adam Smith and the economists (economics has been almost synonymous with liberalism: a creation of it), since the interests of all classes were identical, there was no case for the intervention of the State. But to an age that thinks of the interests of classes as at least in part conflicting, it is the case against intervention that falls. Moreover, it has been pointed out by M. Halévy that there was a fundamental confusion in the mind of the Radicals on this subject. In economics they were against State intervention because they regarded the interests of various classes as mutually complementary. Their juristic doctrine, however, was quite the opposite; in this field they regarded the intervention of the State as necessary in order to reconcile the mutually conflicting interests of society.

There is a further respect in which liberal doctrine has proved, in this age, to have been more fatally at fault; that is, in regard to its rationalist assumptions in politics. Liberalism suffers from what I have called " the rationalist fallacy ". All Liberal thinkers who are at all characteristic, whether Locke or Bentham, Adam Smith or Mill, Helvétius or Constant or Mazzini, Morley or Sir Herbert Samuel, are at one in assuming that politics is concerned with essentially rational beings. The appeal is therefore to reason. How tragically this age has found them out — to its discredit, it is true, rather than theirs! But this mistaken assumption illustrates once more the restrictedness, the individualist world in terms of which they thought. It goes to emphasise once more the theme running all through Mr. Laski's book, that Liberalism historically was the creed thrown up by the middle class in its rise to power.

To be plain, the enfranchisement of the masses has had the effect of debasing the coinage of politics. There is evidence for that on all sides; the vulgarity of current political language, the sentimentalities and half-truths in which it is thought necessary to address a democratic electorate, the lowering of intellectual standards in political discussion. Who can think the standards in contemporary public life equal to those of Gladstone and Disraeli, Salisbury and Chamberlain, Morley, Asquith, Balfour? The appeal to reason is hardly now in case; the masses prefer their thinking to be done for them, and for the rest, to be led. Yet it may be that after the descent to a lower plane of thought and action, that this age has witnessed — the attempt to bring into the foreground of politics the mass of people who have hitherto remained in the background — standards may begin to rise once more as the work of absorption is accomplished. In contemporary Europe, leadership and authority are in the ascendant; when they have had their day, may it not be that the principle of liberty and the appeal to reason will move to the fore again and play their part on a wider field of political action, as once they did on a small one? That at any rate must be our hope.

If there is truth in this hypothesis, and things tend in that direction, it will supply the answer to Mr. Laski's historical scepticism. It will mean that some part of the Liberal spirit — in which this country has given such a rich contribution to the world — will have transcended the conditions to which it owed its rise, to carry over its influence in maintaining the values of personal liberty and of rational discourse in public life into a new age, a new society.

XIV

THE *DÉBÂCLE* OF EUROPEAN LIBERALISM

[1936]

THE popularity of history is one of the most significant characteristics of the intellectual life of our time. It is to be observed not only with the general public, when a biography of Henry VIII or Elizabeth or Cromwell sells into scores of thousands, but with the more selective public of writers and thinkers themselves. For some reason, not only (we may be sure) the vulgar consideration of sales, they feel themselves under the intellectual necessity to consider history, and accordingly take to writing it. In these latter years there has been a number of distinguished recruits to historical writing from other, sometimes remote, fields. There have been scientific romancers like Mr. Wells, no less than eminent politicians such as Mr. Churchill or Trotsky. D. H. Lawrence, if under protest, wrote his history text-book; and even Mr. Shaw, in some ways the most unhistorical of minds, though his contributions to the history of Fabian Socialism are excellent after their kind, at length yielded to the fascination and wrote *St. Joan*. Croce, though primarily a philosopher, has been mingling history with philosophy over the past forty years; and now, the latest convert to the subject, most unexpected of all, is a mathematician and moralist, Bertrand Russell.

There is a profound reason for it: this is an historically-minded age. Everyone knows how this is, or is supposed to be, the age of science. No one has yet observed how much, and more profoundly, it is the age of history, deeply historical in its outlook, impregnated with the idea of historical development, of the ebb and flow of social change; to such an extent indeed

The Débâcle of European Liberalism

that our cast of mind is riddled with scepticism, and in many quarters, notably in contemporary Germany, men seem to have given up the search for truth, defeated by its complexity, disillusioned. Perhaps what sums up more exactly the characteristic outlook of our time is that it is essentially one of historical relativism. (In the appeal to history, on their own interpretations of it, Communists and Conservatives, so dissimilar in all else, are brought together; it is significant that liberalism is excluded from the understanding.)

How different the secure, the certain nineteenth century! However unsettled and puzzled men then felt themselves to be, we, seeing that great age in retrospect, realise what certainties there were, what sure standards of conduct (even if questioned) in public and private life, the lamps of human reason to light a progressive way.

Rationalism, at least with regard to the public affairs of society, the assumption of rational order, even if men were not all equally equipped to apprehend it, the belief that men were open to persuasion as to the right course to follow if only they were sufficiently educated, were characteristic notes of that time. And certainly the conviction gave those who held it confidence in themselves and the enthusiasm to undertake great things for others.

Every man possessed of reason [wrote James Mill] is accustomed to weigh evidence, and to be guided and determined by its preponderance. When various conclusions are, with their evidence, presented with equal care and equal skill, there is a moral certainty, though some few may be misguided, that the greatest number will judge right, and that the greatest force of evidence, wherever it is, will produce the greatest impression.

There is something magnificent about such an avowal. Yet who could believe or say that now? The strength that such conviction gave, the triumphs it achieved, and the defeat in our own time, are the subject of these two remarkable books.[1]

There was much in the conditions of the nineteenth century that made it a favourable time for the reception of such

---

[1] *History of Europe in the Nineteenth Century*, by Benedetto Croce; *Freedom and Organisation in the Nineteenth Century*, by Bertrand Russell.

opinions. (The reception of opinions, a modern would say, is a fact of perhaps greater historical importance than the opinions themselves.) For one thing, these radicals and progressives were not so far wrong in their assumption of rationalism as their liberal descendants have subsequently become; since they in the last century were appealing to a small, instructed, and responsibly-minded audience. They wrote, not even for the whole middle class, but for that small section of upper middle-class opinion, closely associated with the Civil Service and the professions, that has contributed out of all proportion to English intellectual life. James and John Stuart Mill were in the position of civil servants with the East India Company; Bentham, a lawyer rich enough not to practise; Ricardo, a banker; Malthus, a clergyman; and so on. The *Westminster Review*, the organ of the Philosophical Radicals, never had more than a few hundred readers, yet their views and doctrines did so influence the minds of this small but effective class as in this country to mould the political practice of the century.

It was then, both authors are agreed, the age of the rise, the triumph, and defeat of Liberalism. Perhaps one may reflect in passing that neither Lord Russell nor Signor Croce allows quite sufficiently for this factor of political rationalism, that was an essential part of Liberal ideology and was so important in bringing about its downfall. But they are undoubtedly right in making their essential theme the idea of liberty and, what each of them regards as its extension, the principle of nationality. There is a considerable correspondence in their treatment of the subject — all the more notable because they bring very different attitudes of mind to bear upon it. It is these very differences of approach that make the two books, taken together, such fruitful reading.

They have, in the first place, somewhat different scope. Croce's book is, what its title says, a history of Europe in the nineteenth century; it covers the usual ground and in what one is sometimes inclined to think too usual a manner. Russell's treatment is more original; he is not so much concerned to detail what was happening in each country as with what was

The Débâcle of European Liberalism

of significance in thought and in events in the countries of central importance to his theme — Great Britain, the United States, and Germany. The theme is a superb one: it is the dialectical rhythm that ran through the century, of movements towards political freedom met and checked by the growth of industrial organisation.

The purpose of this book [he says] is to trace the opposition and interaction of two main causes of change in the nineteenth century: the belief in *Freedom* which was common to Liberals and Radicals and the necessity for *Organisation* which arose through industrial and scientific technique.

It was the tension between these two that produced much of the richness of that century of crowded history; it was that same dualism, the failure to achieve a form of social organisation which might contain them both, that led to the breakdown, the tragedy of 1914–18.

Lord Russell, one suspects, has been a little inhibited by the necessary technique, so new to him, of writing history: he has not quite launched out into the largeness of his theme, as he might have done *à la longue haleine*. But it is a good fault, and is perhaps a necessary consequence of his approach, which is analytical where Croce is synthetic. As might have been expected, he is at his best in those sections dealing with the history of thought. Two of his best chapters deal with the complex of theories going under the name of Marxism; there is a brilliant exposition and criticism of Dialectical Materialism, that shadow overlying so much of contemporary thought, and of the theory of surplus value, the *bête noire* of so many who attempt the formidable task of reading Marx. Of Marx himself his treatment is kindly and fair, though he has some penetrating suggestions to offer concerning his curious and uncomfortable psychology; but with biography Russell is obviously bored and even a little slipshod — as certainly it is difficult to derive any pleasure from that bleak, jealous, devoted life. The Utilitarians are much more after his heart; he clearly has a great, and deserved, respect for Bentham, who held that everybody acted from the motive of self-interest and proceeded to live a life of complete altruism. One wonders if Bentham ever observed the curious contradiction in himself? Perhaps so, for he was a not unselfconscious man and capable of regarding himself with humour.

In describing the Whig society, in his chapter on "The Aristocracy", that was the background of much Utilitarian and later Liberal speculation, Lord Russell is at an advantage in being able to draw upon the memories and traditions of his family, a good point in an historian, if a new trait in him. For the Whig attitude to the monarchy, for example, there is Lord John Russell's reply to Queen Victoria on being asked if it was true that he held resistance to sovereigns justifiable in some circumstances: " Madam, speaking to a sovereign of the House of Hanover, I think I may say that I do ". Or there is the pleasant story he remembers from his grandmother to illustrate the notorious overcrowding at Holland House dinner-parties, the focus of this Whig society; how on one occasion when an unexpected guest had arrived and Lady Holland called the length of the table, " Make room, my dear ", Lord Holland replied, " I shall have to make it, for it does not exist ".

In America, the section on which is the best in the book, his sympathies are similarly with Jefferson and what Jeffersonian democracy stood for and desired. He says of him:

> It is his belief in the moral sense and the innate goodness of man that gives the basis for his Liberalism. If every man knows, by means of his conscience, what it is right to do, and if what it is right to do is to do good to others, then it is only necessary for the general happiness that each individual should follow the dictates of his own conscience. . . . For the few laws may be necessary; but in the main liberty is all that is needful for the promotion of human happiness.

But he sees, indeed it is his thesis, that this libertarianism that expressed itself in American political institutions was by its nature incapable of keeping in check the enormous monopolies of economic power, the iron and steel trusts, the oil combines, the finance corporations, which grew up alongside of it. He concludes:

The Débâcle of European Liberalism

To master the forces of modern capitalism is not possible by an amiable go-as-you-please individualism. By fastening this now inadequate philosophy upon American progressives Jefferson unintentionally made the victory of Hamiltonian economics more complete than it need have been. The philosophies of which these two men were the protagonists dominated American life until the year 1933.

He might have gone farther and shown, what Croce seems to realise, that it was the mentality of Liberalism that led to the establishment of these agglomerations of uncontrolled economic power; and these, in turn, have disrupted the political fabric of Liberalism and ruined its hopes of international order.

Croce, for some reason, appears to be under the necessity of arguing that Liberalism was not essentially bound up with *laissez-faire*, that its opposition to Socialist doctrines in the nineteenth century was one of expediency, not on the grounds of principle, rejecting socialisation of the instruments of production only in " given, particular cases ". It is a curiously tendentious argument that these pages contain on the relations between Liberalism and Socialism. One need only comment that the " particular cases " where Liberals found it necessary to combat Socialism were in fact fairly general; at almost every point Liberals were to be found combating Socialist influences, whether in thought or in action. The Philosophical Radicals, for instance, hated the teachings of Hodgskin and Owen; Russell quotes a letter from James Mill to Place about a working-class deputation that had been preaching Socialism to the editor of the *Morning Chronicle*:

> Their notions about property look ugly; they not only desire that it should have nothing to do with representation, which is true, though not a truth for the present time, as they ought to see, but they seem to think that it should not exist, and that the existence of it is an evil to them. Rascals, I have no doubt, are at work among them.

Later, Mill wrote to Brougham:

> These opinions, if they were to spread, would be the subversion of civilised society; worse than the overwhelming deluge of Huns and Tartars.

The End of an Epoch

Or there was the opposition of Whigs and Radicals to factory legislation; the *Economist* even opposed the Public Health Act of 1848, which was the result of the appalling sanitary conditions revealed by the Commission a few years before, on the ground that it was an unwarrantable interference with " suffering and evil ", that are

> nature's admonitions; they cannot be got rid of; and the impotent attempts of benevolence to banish them from the world by legislation, before benevolence has learned their object and their end, have always been productive of more harm than good.

Or, again, there was the antagonism of the Anti-Corn Law League to the Chartists, the hostility of Cobden and Bright to independent working-class action in politics. After the middle class had obtained a measure of enfranchisement by the Reform Bill of 1832 they were by no means anxious to share representation with the working classes, whose support they had used in demonstrations like the Bristol riots, as so much leverage to achieve their end. While abroad, in the Liberal movements in France and Germany, the same motives are clearly discernible: middle-class Liberals were willing enough to use the discontent of the masses turning to Socialism in their attack upon the old order and its monopoly of power, but once they had gained a footing inside they allied themselves with it to maintain a new and stronger, because more extended, monopoly of power. It was this that paralysed and kept rigid the July monarchy, at one time so promising and hopeful, and led to its collapse in 1848; while the repression of the working class in the June days was far more savage than anything for which the old order was responsible. Ultimately it destroyed the chances of Liberalism in France for two generations and led to the disastrous Imperialism of the Second Empire. As Tocqueville said of the *Tiers État* to Nassau Senior with his usual penetration of mind:

> Their insane fear of Socialism throws them headlong into the arms of despotism. As in Prussia, as in Hungary, as in Austria, as in Italy, so in France, the democrats have served the cause of the absolutists. . . . Now that the weakness of the Red party has been

proved, now that 10,000 of those who are supposed to be its most active members are to be sent to die of hunger and marsh fever in Cayenne, the people will regret the price at which their visionary enemy has been put down.

It was not until Liberalism was near the end of its time, when it was already irremediably challenged by the growth of Socialism among the working class, that it attempted to recover its position by becoming more " social " in its attitude. By then it was too late. In England in the 'seventies the Conservatives had been more friendly to the Trade Unions and it was a Conservative Government that gave them their legal status when Gladstone had refused. Moreover, when the Liberal Government of 1906 came forward with its social legislation it was taken over from Bismarck, whose " Socialism " was motivated by the desire to check working-class political activity. But the period 1906–14 was, in the nature of things, no more than a St. Martin's summer for Liberalism; and the moral of it reinforced that of the preceding century, that it could only be strong in association with the working classes; when it antagonised them it delivered itself into the power of the old order, still strong, still entrenched.

The latter was at its strongest in Germany; and it was here that Liberals were at their most ineffective, the great majority of them, after the failures of 1848–51, having been seduced by the success of Bismarck's *Machtpolitik* into supporting and applauding the policy of extreme nationalism, themselves making the case for the absolute State and strengthening its autocracy. It is clear that both Russell and Croce regard Germany as the great betrayer of the principle of liberty in the nineteenth century; indeed Croce several times returns to the subject, well acquainted as he is with German thought. He indicts Bismarck, who, once his own power was firmly established, frustrated the training of a politically responsible class, with the result, as Croce pointedly reflects, that the Empire was left with the sort of men portrayed in Bülow's *Memoirs* to govern it. Again, the alliance of the Monarchy, the Conservatives, and the Catholic Centre was too strong for Liberals and Democrats to make headway against; while Croce taxes the pompous and silly German professoriate for the hopeless unreality of their outlook on politics and the failure of academic life to provide any corrective to undesirable tendencies in German society or any standards by which they might be tested or effectively criticised. He returns again and again, at every stage in the history of the last century, and with well-merited scorn, to this subject. There is the German historian who extolled the Prussian law of September 3, 1814, establishing compulsory military service as "one of the legislative acts that mark an epoch and help to understand what history really consists of" — a turn of phrase that only a German could have written. Of the latter part of the century Croce comments:

The scarcity of political sense in the Germans was noted at this time by Germans themselves, who wondered at this lack amid the excellence of all the rest. . . . But there was no end of savants and professors, with that peculiar air of limitation and ingenuousness and often of credulity and puerility in judging practical and public things which is the characteristic of their intellect and mode of life; they were delighted with the strong words and gestures of Bismarck, and the " *Oderint dum metuant* ", the " We Germans fear God and naught else in the world ", of the speech of 1888. . . . Their literature abounded in theories concerning the State, in contrast with the frugality in this respect of the English and the poverty of the Americans, who, as Bryce wrote, had no use for theories on the subject but were satisfied with founding their constitutional ideas on law and history.

It is clear that Croce has not much respect for German interpretations of the idea of liberty, nor for their ways of expressing it in the institutions of the State. His sympathies are unmistakably with those Western peoples, particularly the English and the French, whose political experience has been creative of what is best in the idea of liberty, in the conditions of the modern world. And when we turn to the careful and excellent analysis that Russell gives us of the thought of Fichte and Treitschke we see how right he is. Of that corporative discipline of the will, which he regarded as the root of national education, Fichte wrote:

The Débâcle of European Liberalism

The new education must consist essentially in this, that it completely destroys freedom of will in the soil which it undertakes to cultivate, and produces on the contrary strict necessity in the decisions of the will, the opposite being impossible. Such a will can henceforth be relied on with confidence and certainty.

We know where this leads. After this we are not surprised to hear that

only the German — the original man, who has not become dead in an arbitrary organisation — really has a people and is entitled to count on one, and he alone is capable of real and rational love of his nation;

nor that " to have character and to be German undoubtedly mean the same ". Nor can the German State be hampered by mere pacifism : peace is the ideal of those who love material comfort; the German State has " a higher object than the usual one of maintaining peace, property, personal freedom and the life and well-being of all ". As if peace and liberty, as Croce well sees, do not necessitate much harder and more complex struggle, constant effort, and much wakefulness. How inferior is this German thought to the superb and flaming plea of Dante for universal peace in the *De Monarchia* !

But these traits in German thought and life go back a long way; and the tragedy of modern Europe is that Bismarck should have brought them success in their own crude and forceful way. There is a strong historical case for saying that the total consequence of the irruption of Bismarck upon the European scene was to put back the hope of international order by fifty years and more. Lord Russell, though he recognises the greatness of the man and regards him not unsympathetically, sees clearly that his legacy was a deterioration of the European order. He regards the two figures that were formative, alas, of the modern world, as being Bismarck and Rockefeller : Bismarck, who carried through the national organisation of the State, to the destruction of Liberalism and the exclusion of all else; Rockefeller, whom he takes as the symbol of the organisation of economic power, that has gone far to overwhelm the principle of political liberty which gave it scope and by which it has flourished. He sees the contemporary problem thus, in a world from which Liberalism has all but vanished and in which its remedies have no place:

Organisation to the utmost within the State, freedom without limit in the relations between States. Since organisation increases the power of States, and since their external power is excited by war or the threat of war, increase of merely national organisation can only increase disaster when war occurs. . . . Organisation, with modern industrial and scientific technique, is indispensable; some degree of freedom is a necessary condition of both happiness and progress; but complete anarchy is even more dangerous as between highly organised nations than as between individuals within a nation. The nineteenth century failed because it created no international organisation.

He concludes, therefore, in a sentence, that " It is not by pacifist sentiment, but by world-wide economic organisation, that civilised mankind is to be saved from collective suicide ".

XV

WHAT IS WRONG WITH THE GERMANS

[1937]

THAT there is something wrong with the Germans, something profoundly wrong, the whole of modern European history in the past century is so much evidence. But what it is precisely that is wrong, what is its nature and its roots, is a difficult matter to diagnose. Yet it is exceedingly important that we should understand the issue, for the problem of Germany is the problem of Europe. It is the heart of the problem of how to organise European order, though many English people hardly realise that there is a problem there. They may see quite well the threat to ourselves and the rest of Europe involved by Nazidom, and be prepared to resist its aggression; but they do not see that there is a problem behind that, the nature of the German mind, or at any rate the dominant German mind, for the last hundred years, its essential *difference* from the mind of civilised Europe, north, west, and south.

English people, particularly the more uneducated, are prone to think of Germans as only another sort of English people who happen to talk a different language. The post-war years of a comparatively liberal régime under the Weimar Republic — and how many Germans hated it, groaned under its " oppression " ! — served to encourage the delusion. But it is interesting to observe that it is precisely those people who know Germany least well, who do not know German history or the language or its literature and have never lived there, who entertain most illusions about Germans. English people with a superficial acquaintance with Germany are always impressed by " how much easier

181

it is to get on with Germans than it is with the French; after all, they are our cousins ", etc. But the real truth about that was once put to me by a very intelligent person who knew both well; he said that all the similarities between us and the Germans were on the surface, and all the differences between us and the French; but with a German, the more you got to know him the more you saw how different he was, whereas with a Frenchman, in spite of the differences on the surface, when you really got down to rock-bottom, you realised how like he was, that he accepted the same standards, the same fundamental values. The fact is that the French and the English are part of Western civilisation; it is questionable whether the Germans are. At any rate, there is a problem there, in spite of the great contributions of many Germans to Western civilisation.

The fact is that it is a problem, and a tragic one, within Germany herself: a struggle that has been lit up with a lurid glare for all the world to see in these last years, though the best observers have known that it has been going on for much longer than that, a struggle for the German soul. On one side the forces of reason and culture, of science and the will to cooperate with the rest of Europe, contributing from the resources of German genius to the common stock; on the other, the forces of barbarism, the denial of reason and culture, the cult of violence and aggression, the inflamed inferiority complex, the envy, the jealousy, the *schadenfreude*, the megalomania — in all these things Hitler is the very mirror of the German soul, or the dominant elements of it: hence his success, accompanied by the deliberate cutting themselves off from Western civilisation, which these dominant elements fear because they know it is superior, taking refuge in the depths of their own hideous Teutonism, making of their very worst faults and characteristics a national creed and a doctrine.

Why on earth are they like that? Whatever has made them what they are? One may well wonder. Some historians, going back a long way, put it down to their never having been conquered by Rome. Certainly those parts of Germany that came under Roman rule, the Rhineland, Bavaria, Austria,

What is Wrong with the Germans

have always been the most civilised. But there remained the interior depths of the country, with its barbarian population of Teutons, which — alas for themselves! — retained their integrity, a solid core of Teutonic barbarism unaffected by the influences of civilisation, which we know — despite the ravings of lunatics like Houston Stewart Chamberlain — have all come from the Mediterranean peoples, Greece, Rome, the Jews, Italy, and France. From that solid core of Teutonic resistance, those interior depths, have come a succession of anti-Europeans, leaders whose mission it has been to encourage the Germans, often very heroically, to resist Europe, to assert their own difference, to make a virtue of their barbarism as against reason and civilisation. Luther was the greatest of all such leaders: he came from those depths, not from the civilised fringes of Germany whence sprang his great opponents, Charles V and Erasmus: he spoke out of the burning welter of unreason and belief within him to the hearts of his Germans, ready as always to respond to that sort of appeal.

Luther has been well described as the typical German, the man who of all others sums up the character of the German people. Hitler is in the direct line of succession to Luther; there are very striking similarities between them and in their role as Germans against Europe. What is more comic is that as one reads Tacitus's account of the Germans under their *führer*, Arminius (I suppose he would have been called Herman), one might be reading about contemporary Germany: there is the same *volksgemeinschaft* with its *volk-und-blutsgefühl*, the principle of *führerschaft*, the same background of intrigue and disunity disguised behind the façade of following the leader, the same herd spirit, the primitive love of fighting for its own sake, the people in arms. What a pity for them that they were not conquered by Rome! Western Europe has owed everything to the Roman expansion; if only Germany had come fully within the pale of Roman civilisation, she would have been perhaps supreme in Europe.

But the Germans have always resisted Europe; the influence of Rome in the shape of Christianity came to them late and incompletely. It came in the person of one of the

The End of an Epoch

greatest of Englishmen — Boniface, the Devonshire-man, slain by the proto-Nazis of the eighth century.

At the crisis of Renaissance and Reformation we again see the same dichotomy in the German soul. On one side Luther, the apostle of blind, unreasoning faith, of thinking with the bowels, of nationalism and force, in short of Germanism; on the other Erasmus, a Low German, born in the ancient civilised Low Countries, the protagonist of reason and moderation, of toleration and reform, of peace and a Christian international order, a true European. It is a matter of great historic significance that the issue on which their famous dispute took place was on the freedom of the will; that Erasmus wrote his plea *De Libero Arbitrio*, while Luther replied with his *De Servo Arbitrio*, taking pleasure in the thought how little freedom there is in the human will, revelling in the insistence upon the necessity for submission to the divine will (whatever that may mean: *he* meant Luther), an ultimate denial of individual volition and individual responsibility. How German; how disgusting! " The Holy Ghost is not a sceptic ", cried Luther in triumph; for Erasmus, with an audience of primitive and foolish people, could hardly reply, " The Holy Ghost is just a delusion ". Luther was very subject to delusions. The Germans being what they are, Luther had not much difficulty in winning. But civilisation was with Erasmus: his is the truly tragic figure.

At the time of the Enlightenment, there is the same dichotomy between the pacific cosmopolitanism of Kant (he had Scottish blood in his veins) and the calculated fraud, treachery, and violence of the Prussian Frederick (whom foolish old Carlyle made his hero). There is the same contrast later between the great European, Goethe, and the loathsome German megalomania of Hegel; between the *blut- und eisenpolitik* of Bismarck, which put back the clock in European politics a hundred years and led ultimately to Europe uniting against Germany and compassing her overthrow in 1914-18, and the internationalism of Marx pointing to the future. In our own time there is the contrast between the hopes of social progress and peaceful, international cooperation raised by the

What is Wrong with the Germans

Weimar Republic, all too ineffective though it was, and the relapse into barbarism of the Nazis with their deliberate terrorism and brutality, their pursuit of power at all costs and without any restraints, their denial of all political morality.

The point is, why are the Germans like that? It is impossible to acquit a nation of all responsibility for the way it is governed and the way its government behaves. It is very unreal to make a complete disjunction, as so many illusionists are doing in this country, between the Nazis and the German people. Those who know Germany know that there is something in the German people that responds to this sort of thing, that likes kicking a man when he is down, that does not mind torture and brutality, that respects success at all costs, thinks breaking your word is a sign of cleverness and aggression a sign that you have Providence on your side, that rejoices at the thought of being on the side of the big battalions, and does not care who goes to the wall. The problem is, why are they like it? What is wrong with them?

There are many reasons, geographical, historical, social, that may be suggested why they are as a people so unsatisfactory, such bad Europeans. (I do not need to point out that I do not mean all Germans individually; we all know some who are good Europeans: I mean the Germans collectively, the dominant impression they give of themselves to the world, of which some of the most intelligent Germans have been painfully conscious.)

In the first place, they are rather a frontierless people, and that has had a profound influence upon their mind. They have never had the precise and perfectly definite frontiers that England has enjoyed throughout much of her history, or France, or Spain, or Sweden, or Italy. They inhabit a block of Central Europe and the North European plain, ill-defined by geographical features though rather cut into by them within. The effect has been to make them emphasise their unity as a *volk* speaking the same language, for the want of the very political unity as a nation that England and France achieved so early. But the effect has been even more profound than that, though more difficult to define: this circumstance

The End of an Epoch

has had its influence in communicating a certain frontierlessness, a lack of precision such as classical peoples like the French and Italians so admirably possess, a haziness such as has made them so difficult to deal with diplomatically, which yet expresses itself by its very nature in the boundlessness of its claims and of which the reverse side is the brutal emphasis, when it does crystallise, upon *macht* that appeals to every German, or at least the dominant German type. All this may be read at large in their diplomatic dealings, not only in the period since 1933 — and it is notable how many Germans have felt happier since then, it has enabled them to give rein to what they really feel: in that sense, as their peculiar friends in this country used to claim, Hitler restored their self-esteem, their pride as Germans once again — not only since 1933, but long before that throughout the whole epoch of Bismarck and William II.

We know very well the historical reasons that are adduced to excuse this attitude, and they certainly do explain a good deal: the facts of German disunity and particularism, the fact that the achievement of a national State was so belated, that when it was achieved it was not through the victory of the more civilised and liberal elements, but by means of the most reactionary, the Junker aristocracy of Prussia with its obedient, submissive peasantry, that makes such good cannon-fodder politically and militarily. (So the Nazis have found, too.) It is just as if England had achieved political unity under the aegis of reactionary, fighting Ulster, an Ulster five times as large and ten times more powerful, that extended in a bulk across the whole North of England and into the Midlands, instead of having been united centuries before under the civilised South. One can see with sympathy the plight of the more civilised Germans of the Rhine, the northern seaboard with its cities and their tradition of liberal culture, and the Catholic south with its heart in Vienna. But, all the same, it is hard to forgive the total ineffectiveness, the spinelessness, the disastrous disunity of the liberal elements in Germany. They completely failed to take their opportunity in 1848; Bismarck and the Prussian militarists took it for

What is Wrong with the Germans

them. It has been the central tragedy of modern Europe that the unification of Germany came about that way: a liberal, federal Germany, peaceful and cooperative, would not have collected around itself a world of enemies, brought about by the constant threats to their independence, would not have reduced Europe to a theatre of internecine conflict whose only arbitrament is war, with Germany taking delight in having reduced us to such barbarism, making a cult of it when they should be repenting their sin against civilisation in sackcloth and ashes — and no doubt in time they will receive their punishment as once before they did. But Bismarck succeeded where German liberalism failed; and the latter has failed signally again in our time. I know there were external factors aiding the process, but the German people could have made a better stand for the Republic if they had wanted to, whereas they gave the game to the Nazis. And behold now! — it looks as if Europe will have to go in and civilise the Germans, if we are to have any lasting peace in the world: that is at root what we are fighting for.

Whatever the historical and social causes — and this is no place to traverse the whole of German history — there can be no doubt about the upshot, the problem of the dominant German mind, its character and the threat it holds for European civilisation now. It has been uppermost now for over a century, since the romantic reaction against the cosmopolitan rationalism of the French Revolution. Santayana, in a brilliant diagnosis of the intellectual situation published in the last war, traced its roots back to the tradition of German philosophical idealism, from Kant and Fichte to the magnificent and monstrous megalomania of Hegel, with whom it reached its fullest and most umbrageous flower. And certainly no country can show such a tradition of thought that has so consistently distorted the evidence of common-sense experience, insisted that the world we know is not what we think, erected another world of German egotism writ large in metaphysical terms, and imposed it for philosophy upon the world. It is certainly an astonishing achievement; Santayana diagnosed the truth about the situation twenty-five years ago in

The End of an Epoch

a magnificent passage that does not need a word of alteration but only underlining today:

The transcendental theory of a world merely imagined by the ego, and the will that deems itself absolute, are certainly desperate delusions; but not more desperate or deluded than many another system that millions have been brought to accept. The thing bears all the marks of a new religion. The fact that the established religions of Germany are still forms of Christianity may obscure the explicit and heathen character of the new faith: it passes for a somewhat faded explanation, or for the creed of a few extremists, when in reality it dominates the judgment and conduct of the nation. No religious tyranny could be more complete. It has its prophets in the great philosophers and historians of the last century; its high priests and pharisees in the Government and the professors; its faithful flock in the disciplined mass of the nation; its heretics in the Socialists; its dupes in the Catholics and Liberals, to both of whom the national creed, if they understood it, would be an abomination; it has its martyrs now by the million, and its victims among unbelievers are even more numerous, for its victims, in some degree, are all men.[1]

No country in the world has had such a tradition of thinkers who believed only in the assertion of the will as against reason and common sense, in the supremacy of force, the desirability of war, the State as the *terminus ad quem* of all politics, the be-all and end-all of all social endeavour, the denial of freedom and international order, the futility and even the unmorality of international peace: Fichte, Herder, Arndt, Schlegel, Hegel, Treitschke, Clausewitz, Nietzsche, Houston Chamberlain, Bernhardi, Spengler, and the still lower and coarser vulgarisers of the tradition among the Nazis, Rosenberg, Hitler, Moeller van den Bruck, Goebbels. The strength of the Nazi *weltanschauung* is due — English people do not realise it, for they regard it, quite rightly, as a mixture of mystical nonsense and calculated lies — to its being the culmination of a long tradition going on in Germany for over a century. Nazism is but the contemporary version of that tradition, coarsened and jazzed-up to appeal to the total lack of any cultural standards

---

[1] *v.* Santayana, *Egotism in German Philosophy* (published 1916, republished 1939), p. 69.

What is Wrong with the Germans

in the lower middle class. It is this tradition translated into the language of Hollywood.

Inconceivable as it is that any civilised person should fall for it, yet the Germans do: it is no use blinking the fact that the Nazi creed and code answer to something deep down in the German nature — it would not else be so successful. The fact is that the Germans are a singularly credulous and gullible people, without any critical standards, so that they can never tell a good egg from a bad egg intellectually, whether in politics, or art, or life. No other people in Europe would have believed the inspissated, congealed nonsense first of Fichte, then of Hegel, then of Treitschke and Houston Stewart Chamberlain, then of Spengler, now of the Nazis. The intelligence of the French, the common sense of the English, the scepticism of the Italians, would never have stood it. Yet the Germans lap it up, and the more pretentious and pseudo-academic it is, the more they like it. Hegel was undeniably a genius, though a first-class disaster to European thought; but nobody except the Germans could have swallowed the pretentious rubbish of Houston Chamberlain's *Foundations of the Nineteenth Century*. The book was intended to prove, with a wealth of irrelevant references and misplaced ransacking of books good, bad, and indifferent, that the culture of Europe was the work of *die Germanen*, *i.e.* the Teutons, in which term Chamberlain was good enough to include the French, the Celts, the Latins, and even the Slavs. If he meant to say that European culture was the work of Europeans, he did not need two volumes to say it in. If he meant to say that European culture was the work of the Northern-European peoples, he was plainly wrong, for it owes its whole foundation to the Mediterranean peoples. But the Germans must have their pernicious rubbish or their pretentious nonsense in two fat volumes: there are the pseudo-philosophisings of Keyserling, the *Travel Diary of a Philosopher* (in two volumes); the notorious *Untergang des Abendlandes* of Spengler (another two large volumes), of which it is only too easy to read the motive, a piece of *schadenfreude* erected after the German manner into a whole system of thought: since Germany was defeated in the last war, *ipso facto* Western Europe was declining to its fall. So naïve, so simple; yet all elaborated and systematised *ad nauseam* in the typical German manner. So also with the even coarser and more brutalised doctrines of Rosenberg and Hitler: the former's *Mythus des 20ten Jahrhunderts*, fat and foolish and indigestible, while *Mein Kampf* also has its two volumes.

Why are the Germans such fools as to swallow it all, or even in part? The answer is given by Hitler himself in a celebrated sentence which he removed from the text of *Mein Kampf* as he drew nearer to power: " German political leaders have no conception of the extent to which the German people have to be gulled in order to be led ". He understands his Germans well; and indeed, it must be said to his credit that nowhere does he disguise his contempt for them or his low estimation of their intelligence. *" Die Intelligenz der gewöhnlichen Leute ist sehr beschränkt und dafür ist die Vergesslichkeit gross"*, etc., again and again. He has certainly proceeded to act in accordance with his estimation of them, and he does not seem to have been mistaken. Nor is it a new feature in their character. Before they lapped up the pernicious Nazi rubbish, they lapped up the Hohenzollern claptrap: so much so that it took four years of war, with all the nations against them, before they broke. For the Germans are an obstinate and — no one will deny — a physically courageous people. If only they had less physical, and more intellectual, courage, Europe and Germany herself would be the better and the happier. But perhaps, as Santayana says, they do not in their hearts want to be happy: according to Fichte, self-assertion, not material success, is the goal. He elaborated an educational system so thoroughgoing that men should be incapable of willing anything but what the State wills them to will. The Nazis have brought that into being: he was an exact precursor of their ideas: they are very deep-rooted in the German mind.

The greatest asset in the inculcation of these savage and retrograde ideas in the mind of a passive people is their submissiveness, their readiness to obey, their lust to be led. They are an excessively obedient, docile people: as someone has said, the only great people in Europe who have never had a

What is Wrong with the Germans

Revolution, at any rate a real Revolution. Penetrating observers both without and within have noted this fatal trait in the German people.

They will never rise [wrote Bakunin]. They would rather die than rebel ... perhaps even a German, when he has been driven to absolute despair, will cease to argue, but it needs a colossal amount of unspeakable oppression, insult, injustice, and suffering to reduce him to that state.

Nietzsche observed from within:

They are always so badly deceived because they try to find a deceiver. If only they have a heady wine for the senses they will put up with bad bread. Intoxication means more to them than nourishment; that is the hook they will always bite on. A popular leader must hold up before them the prospect of conquests and splendour, then he will be believed. They always obey, and will do more than obey provided they can get intoxicated in the process.

How prophetic and true this is. When Bismarck made it clear that he did not want any more conquests, he became comparatively unpopular; so was William II for the first few years when he tried to carry out a good-neighbour policy, fairly pacific in intention. It cannot be denied that his glittering, competitive rodomontade after 1892 appealed to the Germans, their naïveté, their childishness, their self-importance, their envy of others with a more secure place in the sun. After a whole essay on the subject of the jealousy the English felt for the Germans, Ernst Moritz Arndt concluded naïvely, " If only we were where you are now! " A sober-minded man like Stresemann did not really appeal to the Germans, nationalist as he was; it needed somebody like Hitler — who is in all his moods a medium of the German people, a mirror to express their rasping inferiority complex, their envy, their brutality, their unreliability, the simple belief of peasant cunning that it is clever to lie, to go back on your word — it needed a Hitler to arouse the German people. They like that sort of thing, or at any rate a certain side of them does, the dominant side. Then it is idle to deny that they are responsible and to put off their misdeeds on to their leaders.

The End of an Epoch

But why are they like this? Is it that they cannot bear the sober reality of the world, as the French or the English or the Americans or the Scandinavians see it? There is a great deal in this: they feel uncomfortable with a plain, common-sense view of the world and life: it revolts them. Their characteristic writers all animadvert against our lack of idealism, our low commercialism, our want of chivalrousness. Why cannot they bear the plain truth? Why must they assert themselves against the facts of life, the very conditions of existence? The answer was once given me by a very distinguished German, who confessed that they feared that if they did not they would go under; that they had to assert themselves or they felt null and void, insufficiently alive, just nothing at all; that at the heart of the German people there is a real neurosis.

Of that neurosis we are all, as Santayana says, now in some degree victims. All these characteristics of the German mind, the inverted sentimentalism the reverse side of which is their brutality, the unsureness of themselves that expresses itself in their disgusting exhibitions of aggressiveness towards others, their persecution of minorities, their erection of anti-Semitism into a deliberate policy — it has been called the " socialism of the lower middle class ", but it is rather a pathological disease, a crime against civilisation: all these things are very understandable psychologically. They are symptoms of a dominant neurosis verging upon madness.

But one does not cure a homicidal maniac by letting him have his way. One shuts him up for the safety and security of the rest of the community. At the very least, our estimation of the German mentality must influence our conception of our war aims and the kind of peace that is safest for Europe. A mentality such as has been dominant in Germany over the past century does not bear trifling with; it is no good trying to compromise with it; the only safe thing is to root it out.

XVI

GERMANY: THE PROBLEM OF EUROPE

[1941]

(1)

JUST before the collapse of France, one of the greatest authorities in France on the German mind — Professor Edmond Vermeil of the Sorbonne — published an admirable book on this most important of subjects for us all. Naturally in the events that followed, his book was overlooked here, and is now overlaid by the black-out that has descended upon France's intellectual life.

Two years ago Professor Vermeil wrote what is far and away the best account anywhere of the whirlpool of political discussion and imagining — for you can hardly call it thought — into which Germany has plunged in the past twenty years: *Doctrinaires de la Révolution allemande, 1918–1938*.[1] Now he has followed it up with a magnificent study, *Germany: an Essay in Explanation*. It certainly needs a good deal of explanation; and it gets it, more fully and with greater knowledge and penetration than anywhere else I know.

I do not think that we have had any similar attempt in English to understand the meaning of German history, what it implies for Europe, to what it has led the German people, the kind of character and mentality their past history has produced in them, the problem of their place in Europe — for, in short, Germany is the problem of Europe. This book should certainly be translated; in the absence of a translation I will do my best to give some idea of what it says, even if I fail to do justice to its excellence.[2]

Vermeil shows, as the result of his study of the evolution of

[1] v. p. 199.   [2] It has since been translated as *Germany's Three Reichs*.

The End of an Epoch

the German people, where their restless desire, amounting to a lust, to impose their domination over the rest of Europe springs from. To attempt European hegemony is, indeed, in the obvious logic of their modern history; but he traces its springs back, paradoxically but convincingly, to the failure of the Germans to achieve a normal statehood, such as Western Europe, England, France, Spain, attained as early as the fifteenth century.

Germany remained right up till the nineteenth a mosaic of small States and princedoms, without real unity, without the effective institutions or the national consciousness of a modern State. That has had all sorts of consequences for their political life. It threw them back upon the dream of universal domination, both material and spiritual, that has haunted their inner mind through the ages.

Everybody knows how the dreams of a revived Holy Roman Empire of the German People float in the megalomaniac minds of Hitler, Rosenberg, and their following; and how this year they have had a temporary quasi-realisation in the subjugation of France, with Italy subordinate, as in the Middle Ages. But like the dream it is, it will not last long. The real lesson of European history is quite different: *it is that no one Power is strong enough to hold all the rest in subjection*, and that therefore we can only build a United States of Europe upon the basis of cooperation with each other. What is interesting is the light that such a dream, and the tenacity with which it is pursued, throws upon the inner nature of Germany. Also very important to understand if we are to deal successfully with her in the future — not idiotically as in the past twenty years — and see her put in her proper place in the European frame.

All through her history Germany has displayed a strong undercurrent of resentment, resistance, jealousy, *schadenfreude* against the West. Perhaps it may be due to what Freud calls a love-hate complex; but the consequences are very uncomfortable for us all, and it must be put an end to this time once and for all. In her early history Germany resented her conquest by Christianity, and killed St. Boniface, the great Englishman who brought her into the fold. Later she resisted and resented

Germany : The Problem of Europe

the spirit of Rome, the spirit of international order and universalism. In modern times she has hated and fought French rationalism, English liberalism, and humanitarianism. These three forces, Roman, French, English, have been the great civilising influences of the modern world.

In spite of herself, Germany has undergone their influence in part, and created great things out of them in consequence — the philosophies of Leibniz and Kant, the art of Bach and Goethe and Beethoven, German science. On the other hand, the reaction has been tremendous, and in modern times seems to have been getting ever stronger.

As against the West, through which European civilisation came to them, they have made titanic efforts to build up their own alternative, and not content with that, to make it prevail. Their first great effort in this direction on the stage of world history was with Luther's revolt against Rome, his destruction of Catholic unity, a movement so comparable to the Nazis and Hitler's destruction of our international order. Vermeil points out that what was involved in the conflict between Erasmus and Luther was the spirit of universal Christianity, in Erasmus's interpretation moderate, tolerant, reasonable, against Luther's national religion, irrational, fervid, egoistic, insisting upon blind faith and obedience. How well we recognise the symptoms and the choice; and what a mistake the English nineteenth century made to crack up a hooligan of genius such as Luther was.

Keyserling has pointed out — it is very revealing — that the characteristic of the West is the belief in Man, in the ultimate value of human personality and the desirability of extending his control over his own destiny through his reason. As against this, the German alternative, as it has unfolded itself, is to follow the promptings of dynamic and aggressive nature, the blind faith of either personal mysticism or collective delusion, the insistence upon the Will as against Reason.

" We want so to educate the citizens of the State ", said the philosopher Fichte, " that they shall be incapable of willing anything other than the State wills them to will." There is the whole Nazi philosophy of education in a nutshell a century and a half ago.

This over-emphasis upon the Will in German thought has been wonderfully brought out by Santayana in his *Egotism in German Philosophy*. He too shows, like Vermeil, that a fundamental characteristic of the German mind is an inexact sense of reality. They see the world in the light of their own dream, their own egoism, their own nightmare, one might say, and then by the most terrific efforts of the will impose that nightmare upon the world. It is terrible to think that even though they assuredly will not win, the Germans have for a time imposed their nightmare upon us all and reduced the world to their dream of it.

Vermeil goes on to show how all this has worked out in later German history. Their belated unification, and the immense successes they had in the nineteenth century, quite turned their heads. It is ironical to think that the nation that has come nearest to achieving European hegemony is that one of the Great Powers least fitted to exercise it. Far better Louis XIV, or Napoleon, or Philip II: they were at least civilised. The total result of this frantic pursuit of power, of this double-quick reaching-out by an immensely strong but completely immature people for universal power, is to be seen in the modern degeneration of Germany and German standards, politically, socially, aesthetically, spiritually.

Bismarck's achievement of German unity, and German ascendancy in Europe, was a great piece of work, and he had the moderation of an aristocrat and a cynic, which enabled him to maintain it with success for a time. But at heart there was a void: the Bismarckian Empire stood for no fruitful, liberating idea such as, say, the French or American Revolutions did, or the October Revolution in our own time. At heart there was just nothing: but force, political mechanics, cynicism, intrigue, treachery, blackmail. No wonder contemporary Europe was alarmed by the apparition of the new Power on the European stage. And with the great man who worked the machine gone, its final nullity was revealed in all the horror of the Wilhelmian era.

Germany: The Problem of Europe

Under William II vulgarity reigned, conceit, self-assertion, aggressiveness, flattery: the picture of that court revealed in Bülow's *Memoirs*, the falsity, the Byzantinism, the treachery — what a commentary they are on German public life! To think that a great country — the keystone of Europe — should have been ruled by a lot of neurotics and megalomaniacs, like William and Bülow, and the pathological Holstein. As Vermeil says, it was utter bankruptcy, political, moral, aesthetic, spiritual. The nemesis of it all was the war of 1914–18.

But the situation has only gone from bad to worse with the Nazis. Germany was not then, at any rate, ruled by criminals. The nation-wide propagation of a pernicious and false racialism, in place of education, had only begun, under William II, to reach what heights with the Nazis we well know. The point is that the Nazis are the inheritors of this immensely powerful tradition inside Germany: hence their strength and success: they represent what so many Germans believe, and others like, if they do not believe. As Vermeil says, *Nazi doctrine is the degeneration and vulgarisation of the intellectual tradition of Germany*.

We do not need to go again here into his clear demonstration of Nazism as the creation and instrument of the German ruling cliques, particularly the militarist circles from which it emerged and with which it never lost contact. Hitler was their missionary to the people, his movement their necessary liaison with the masses, with all the cynical and hideous preachings of anti-Semitism, anti-Communism, anti-Freemasonry, a bogus anti-Capitalism even, to seduce them.

Germany, he says, has destroyed the European order, not for a concrete plan but simply from instinctive impulse. The Germans refuse to see that in the British Empire there is a " real model of an empire, material and spiritual, which knows how to leave to humanity its liberty ": a very remarkable tribute coming, as it does, from a Frenchman, of all the more value since it comes from outside, though it should be obvious enough that it represents no more than the truth. As against that, what have the Germans to offer? Nothing but the cult of aggression, force for its own sake, devoid of all scruple or moral value, an ultimate cynicism that will prove as exhausting to the German people as it is disgusting to all who watch its course.

The German people, said Hegel, have the right to impose their domination by *force* and *fraud*. " We are a people ", wrote Moeller van den Bruck, the elect thinker of the Nazis, " who are destined never to leave others in repose." Either world-domination or total downfall. *Weltmacht oder Niedergang* is the terrible dilemma that the Nazis have brought the German people up against.

Vermeil ends his book with a most valuable psychological analysis. One of their well-known characteristics is their inability to recognise any responsibility for their acts. John Dove, when he travelled in Germany after the last war, noted that he never saw any sign of regret for the tragedy into which they had plunged themselves and Europe. It must be brought home to them properly next time.

Vermeil remarks that it will not be enough to deal merely with the Nazis: what is necessary is the eradication of all the nuclei of the Prussian military tradition, from the army, political and social life, intellectually, and in the education of the people. That must be taken literally: it needs to be rooted out. The roots and ganglia of the German outlook are so deep and widespread that Europe will have its work cut out, merely on the intellectual side, after the war is over. We cannot allow the Germans to make a European war every twenty-five years to prove that the last one was wrongly decided against them. It is our role to organise the whole of Europe to see that it is done properly, for that is the absolute *sine qua non* of European peace and of an effective working international order.

(II) [1939]

The problem of Germany is the problem — almost the despair — of Europe. There needs no apology for returning to it again and again, considering it now from this angle, now from that, since we have it with us always. If Germany involves Europe in another war, and is defeated, it will still be there. What are we to think of it? There is every reason to

Germany: The Problem of Europe

inform ourselves of what contemporary Germans are thinking — of what passes with them under the guise of " thought ".

How much we have lost already by not taking advantage, in time, of the information open to us all in such works as *Mein Kampf*, in the writings of Rosenberg and Goebbels, and the whole tradition of militant nationalism and racial mania, of which the Nazi régime is the culmination.

This French book [1] offers us a very fair chance of making up for lost time. It is the best survey I know of the thinkers who have made the thought of the Germany we have to deal with : of Rathenau, Keyserling, Thomas Mann ; of Spengler, Moeller van den Bruck, and the group of writers in *Die Tat*; of Hitler, Rosenberg, Goebbels, and all the lesser fry among the Nazis. Professor Vermeil is a distinguished authority at the Sorbonne on contemporary Germany and German thought ; and his book has all the virtues of the French academic mind at its best, precision and lucidity, careful documentation and a sound instinct for what is sense and what is nonsense. Nor, though it is critical, can it be said to be unsympathetic ; for one thing, I should not be inclined to regard the thought of the idealists, Rathenau, Keyserling, Mann, with such respect as he does, to regard them as having a place in the " great German intellectual tradition of the eighteenth and nineteenth centuries ". But then I have always considered German thought very much overrated in this country since the later nineteenth century ; we have a very much sounder and more sensible tradition ourselves in English empiricism, in Locke, Hume, Mill ; so perhaps that evens out. This book certainly ought to be translated, along with the more important works on which it comments — Moeller van den Bruck's and Rosenberg's for instance — so that we shall know what these people think even when it is dangerous nonsense.

What is remarkable is the continuity of thought, the strongly marked common characteristics among these writers, in spite of their different positions and associations. " Writers or publicists, orators or men of action," M. Vermeil comments, " the thinkers here dealt with all go back to the traditions

[1] *Doctrinaires de la Révolution allemande, 1918–1938*, by Edmond Vermeil.

of the nineteenth century. Since the end of the Holy Roman Empire, the intellectual *élite* of Germany has never ceased meditating upon the origins, the nature, and destiny of the German Reich." They have a great deal more in common than this: almost all of them, whether Liberal or Nazi, idealist or realist to the point of cynicism, share a distrust and dislike of the West, of England and France, and what they stand for in European civilisation, amounting in some cases to horror, but always to active hostility. One knew before that the fatal and systematic Spengler, who constructed, as Germans will, a whole sociology out of his own *schadenfreude*, hated England and believed in an inevitable conflict between England and Prussia, of which 1914 was only the beginning. It is interesting to note how many of these thinkers, particularly among the Nazis, subscribe to his view that the War of 1914 was only the first phase in the great German Revolution — and as such justified. On that view, the abnegation of any responsibility for the War becomes somewhat superfluous. And, in any case, so many of these thinkers regard war as a good thing in itself, and life without war with horror.

But it comes with something of a surprise to realise to what an extent a so-called liberal like Thomas Mann, in his heyday, shared the general German hostility to the West, the " eternal German protest " — as M. Vermeil calls it — against the intellect, against the life of reason, his criticism of our over-civilisation, our decadence and hypocrisy. The world could do with a little more of our over-civilisation at present. It is tragic to observe how thoroughly, with what laborious exactitude, these idealist writers — Rathenau, Keyserling, and Mann — paved the way for the brutality and vulgarisation of the Nazis. Rathenau died their victim; Thomas Mann is now in his old age at leisure to repent, an exile in America.

One does not need to go again into the extraordinary system of thought erected by Spengler on the basis of German defeat (*ergo*, defeat of the West); it has been sufficiently criticised before. And yet there is something so representative of Germany in its completeness, its heavy pretentiousness. Its whole conception of a morphology of cultures is, of course, false, but so typically German. Spengler's conception of a culture is altogether too anthropomorphic, much too rigidly defined, something marked off like one living entity from another. On the one hand we have a " culture " defined as a person, which is born, lives, and dies; on the other we have the person regarded as having no individual existence: an inversion so characteristic of German mental processes. It is all very like Hegel, *fons et origo* of these intellectual ills. The West knows that it is only individual human beings who have a real existence and that groups exist to aid them in attaining a fuller and more satisfactory life. That is the fundamental difference between Europe, particularly Western Europe, where European civilisation originated and in which it is still located, and Germany. It happens that this position is also the true one; the best elements in Germany recognise it. It is a pity that they have been found so weak and ineffective; but Germany will come back to it, leaving this self-willed ostracism which M. Vermeil so well diagnoses in nearly all these writers: " *cet ostracisme de plus en plus radical qui résume toute l'attitude de l'Allemagne contemporaine a l'égard de l'Occident européen* ".

There are other characteristics, which Spengler expresses, that recur again and again in the others. For example, the hatred of reason which " kills life ". But why? one wonders; surely intelligence and clarity of mind enable one to live more satisfactorily? To put it at the lowest, it is the intelligent animals that survive. One wonders why these Germans — all these thinkers bear evidence to it — hate reason so much: is it because they are so bad at it? All this goes along with a pathetic insistence, which " liberals " like Thomas Mann and Keyserling and Rathenau share with brutes like Spengler and Hitler and Rosenberg, in the superiority of German culture. They must be superior or nothing. It is the same cry in the realm of thought as in politics: " Either a world Power or nothing ". Spengler has a whole theory — he would have — that a " culture " must dominate. No conception of collaborating with others to make a more varied, a richer, more fruitful civilisation. Actually it is enough to discern the difference between cultures; one does not have to think all the time in terms of superiority and inferiority, but of the particular contribution that each has to make to the whole that is Europe. To the really enlightened there is a European culture in which German music, where Germany really has been supreme, takes its place along with Italian and French art, Greek and English poetry. Modern science is essentially European.

The fact that music is the art in which Germany is supreme is one of great significance, if one could elicit it. There is something in this magnificent structure they have built up out of sound — this inner world of experience, to which some of their best poetry, for example Rilke, conforms — that reveals their weakness in the realm of external form, in imposing rational order and control upon their experience, where the Latins, and of them the French above all, are so strong. There are moments with all these writers, especially with Rathenau, even with Spengler, and still more surprisingly with the cynical candour of Goebbels, where one breaks through the crust of neurotic assertiveness, the over-emphasis of people not sure of themselves, and one gets a glimpse of the depths of formlessness, of indecision, of extreme relativism, of a scepticism underneath amounting to real nihilism (*pace* Spengler), that have their counterpart in the violence and brutality of the Nazis. Rathenau allowed the truth to appear, in spite of all his hopes of a new order led by Germany, in a passage where he said that beyond the Rhine " there is neither form, nor style, nor any real desire for liberty. Everywhere, on the contrary, feebleness of will, permanent confusion between loyalty and dependence, between autonomy and anarchy, between work and servility. The masses bow before all the forces of the day. *Tout le slavisme inhérent s'étale ici en face d'une poignée de Germains authentiques.*" There is an equally remarkable passage from Thomas Mann, which deepens the despairing picture of a people without form, open to all the winds that blow, exposed to all the contradictions, " *nation qui ne s'enferme jamais dans un réseau solide de systèmes, de morales ou d'institutions. L'Allemagne remet tout en question, inlassablement.*"

How true that is; and they have succeeded in laying everything in contemporary Europe open to question yet once more. What this whole development of post-War German thought has led to we are only too well aware. M. Vermeil devotes the second half of his book to the critical examination of Nazi doctrine. There is no space to go into it in detail; one can only agree that it means a terrible debasement of intellectual standards, the impoverishment and dragooned monotony of German thought; while culturally nothing could be more sterile: it means the laying waste of a great country that should be making its contribution to European civilisation.

(III)

It is extraordinary how little people realise in this country, particularly on the Left, what a chasm divides the German mind and German thought from the mind and thought of this country, and, indeed, of Western civilisation. It is a very dangerous ignorance, particularly again for the Left. English people, especially if they are liberal-minded, cannot bear to think that other people may be very different from them, especially, oddly enough, if the others are not liberal-minded. It is noteworthy, too, that most of the liberal illusionism about the Germans and the German mentality, which does duty for knowing and thinking about the subject, comes from people who do not know Germany or German history, the German language or German thought. What is their plain duty is that they should get down to studying the question; it is a question of cardinal importance for us, and for the future of civilisation.

German writers themselves make no bones about the chasm separating them from the West, the kernel of European civilisation. Indeed, they have willed it so. From Herder and Fichte onwards there has been among their leading thinkers a gathering momentum of assertion of their difference, denial of Western standards. First they denied the rationalism and individualism of the eighteenth century; French classicism and English empiricism. Then Fichte, Adam Müller, List denied the cosmopolitanism, the universalism of Adam Smith and the

English economists. They asserted the virtues, the superiority of irrationalism, of collective feeling, *Volksgefühl*; they insisted upon the national State, based upon racial community or affiliation, as the real functioning unit in economic, as indeed in most other, matters. They refuted nineteenth-century aspirations for peace, and asserted the superior virtue of war, without which (*pace* Arndt, Clausewitz, Hegel, even Ranke, let alone Bernhardi, Treitschke, Nietzsche, Spengler, etc.) the world would sink into materialism. They came to pursue dynamism, the worship of success and power, to attain which all means were held valid, so that force and fraud, violence and treachery were erected into the first principles of State. *There is a difference.* Other thinkers and statesmen in other countries have sometimes thought the same things; but it is in Germany that they have been erected into the first principles of State policy. Hitler goes back by a direct line of affiliation to Bismarck and Frederick the Great. It is idle — besides being dangerous and a disservice to their country and to Europe — for liberal illusionists here to deny this: Hitler is quite conscious of it and is proud of it. It is not surprising that contemporary German thought has come in the end to deny, as Mr. Butler says, the virtue of truth itself, the ultimate value of the human personality, two of the fundamental principles upon which civilisation is based.[1] In the Nazis we see a deliberate denial of all principle, a ruthless moral nihilism, enthroned.

But they are only the culmination of the whole tradition. That is what people do not sufficiently understand in this country. The mentality of militant aggression, with the whole complex of doctrines that goes with it, German superiority, decadence of the English, hatred of rationalism and the West, is not merely the dominant tradition of Germany in the past century; it is *the* German tradition *par excellence*. You would have no idea — it comes as a constant surprise to one — of the extent to which practically all German thinkers have been under the influence of the tradition or have contributed to it. An extremely well-educated and discerning friend of mine, who knows German well, has translated from the language, told me

[1] *The Roots of National Socialism*, by Rohan D'O. Butler.

that until he read Mr. Butler's book he had no idea of the strength and true character of this tradition of thought with the Germans. The truth is that in one form or another they find it irresistible.

Take even the case of Thomas Mann: he is supposed to be the most eminent living representative of the liberal opposition to Nazidom, a champion of civilisation, of the universal spirit. Yet Mr. Butler tells us that Mann's main political work, the *Betrachtungen eines Unpolitischen*, argued that " German culture was not merely hostile to Western civilisation in that it lay behind the Great War. That culture was in its very being the enemy of that civilisation. And that culture was in the whole tradition of Germany." He was, of course, perfectly right, as against fools over here who know nothing about the subject. But, and this is the point, Thomas Mann placed himself — it was 1917 — with German culture against Western civilisation. Since the logical fruits of his choice have made themselves clear in Nazi Germany, he has been driven to take refuge in the remotest West; but, though we have had various pompous political pronouncements from him, I do not know that he has seen fit to repent or to disclaim his *Betrachtungen eines Unpolitischen*.

The fact is that it has not been easy for English readers to acquaint themselves with the body of German thought on politics during the all-important past century, when Germany came to occupy the key position in Europe, so that the German problem became the problem of Europe. We were acquainted with some individual thinkers — Kant, Fichte, Hegel, Treitschke — but not with the amazing force and cohesion of the whole. It is the more remarkable, as Mr. Butler says, " since the dominant German outlook springs from a very bold and imaginative corpus of thought, and one which, almost throughout the course of its growth during the last century and a half, has been intimately interconnected with the politico-social development of the nation ".

English readers will have no such excuse in the future. Mr. Butler's book is one of those rare examples of a book which is, if anything, under-estimated in its publisher's blurb of it.

The End of an Epoch

It is, indeed, " extraordinary that so few English writers have paid serious attention to the pedigree of National Socialism. Mr. Butler's very able book closes this perilous gap in our political literature." It is a very remarkable book; it does more. Its somewhat too topical title would give one to suppose that it was one more work of political journalism about Germany. But not so; it is an admirable work of academic scholarship, very well documented and well written, tracing the whole development over the past 150 years of this dominant German tradition, *the* German tradition in fact. It is to be welcomed as the first book of an author who will be heard of: one of the two or three most promising of our young historians. What is so remarkable in quality is the way this young historian keeps his head : such common sense and so mature a judgment, animated by an uncommon gift for summing up a complicated train of thought concisely, sometimes brilliantly. A grim humour relieves the deplorable subject. But, indeed, what is one to think of such maniacs of genius as Nietzsche, Hölderlin, Moeller van den Bruck, Hitler; such sombre enemies of civilisation as Spengler, Ludendorff, Bernhardi, Treitschke, Clausewitz; the pretentious pomposities (taken so seriously) of Wagner, Rosenberg, Houston Stewart Chamberlain, Schelling, Schlegel? What are we to think of a nation that has put up with and been so influenced by such criminal outpourings, such dangerous nonsense, such vile traditions of anti-Semitism, inflamed inferiority complex, and *schadenfreude*?

(IV)

The question of Germany is the core of Europe's fate, and that is a very important matter for the whole world. If we do not understand it and deal properly with it, there is no hope of peace : we shall have a third German war in our time. That is reason enough for recurring to the subject, however much many of us may be bored by it — for Germany is nothing so much as the world's bore — the lamentable truth is that there are few enough people in this country who appreciate the extent of the difference between Germany and civilisation, the reasons in her history and character for her pathological condition, the doubt that remains whether she will still have learned her lesson after all the suffering she has inflicted upon others and brought upon herself.

There is the danger; and there is nothing more dangerous for our future than the combination of English good-nature and ignorance on the subject. The English cannot bring themselves to believe that there can be really evil people in the world. It is not for nothing that Pelagianism is our one characteristic heresy (Pelagius, if not an Englishman, was at any rate a Briton).

Thomas Mann, recipient of a Nobel prize, is regarded as the foremost living German writer. As a writer he is overrated, both by himself — he treats himself as a classic — and by others. But that does not make him any the less valuable as evidence of the German condition and commentator on Germany's course. Born of an old Lübeck burgher family, he is at any rate *very* German. This book consists of essays and addresses, directed almost wholly to the German problem, and commenting on the course of events in the last twenty years.[1] Whatever we may think of it as literature — and like most German writing it is heavy, flat-footed, humourless, pretentious, without lightness or brightness — yet it is important to our understanding the true case of Germany that we should stick it out and read it.

The book begins with a plea for the German Republic, uttered in Berlin in 1923; a very tame, oblique, and ineffectual affair. No wonder the Republic fell down when it was supported in such a half-hearted manner. But its importance is that it was in some sense a renunciation of Thomas Mann's stand in the last war, when like the great body of German intellectuals he welcomed the war, wrote lyrically about it (so long as all went well), and underlined the contrast — all too evident indeed — between Germany and Western civilisation. Though Mann has never been able to make a clean breast of what he did write in the last war and unhesitatingly repent of it, it is to his credit that he has never

[1] *Order of the Day: Political Essays and Speeches*, by Thomas Mann.

gone back since 1923 but has gone on to harden his opposition to all that has made Germany a criminal among nations.

He knows, if many of our sentimentalists do not, that National Socialism is but " the poisonous perversion of ideas which have a long history in German intellectual life ". There is the difficulty and the danger. Of course, the Nazi bestiality would never have had such a hold on Germany if it did not appeal to something very deep in the German people. Of course, Thomas Mann, like every other good German, is prepared to allot a share of responsibility to other nations for the rise of the Nazi tyranny. But that is only in keeping with the fundamental trait among Germans — their inability to face responsibility for what they do. *They* never made the last war; *they* never made this; everybody else is responsible except them. In the year in which I lived in different parts of Germany after the last war, I never heard a word of regret for the war; while the English were beginning to sit superfluously in sackcloth and ashes for a war which they did not begin, had never wanted, and were not prepared for. Mann is well aware of this pathological lack of any sense of responsibility among Germans — and yet, in one place, himself saddles us with the responsibility for Bismarck's Germany, " which came into being through England's benevolent neutrality ". So far does this trait go! Mann sees now that only a terrible and final lesson can have any hope of bringing Germany to her senses, and yet he warns us that even now it is questionable whether the bourgeois German will learn from his experience.

Thomas Mann has learnt since, in 1914, he sang the praises of war, and wrote that if Germany were defeated " Europe would have no peace from German ' militarism ' till Germany stands again where she stood before this war; on the contrary, only the victory of Germany will guarantee the peace of Europe ". If this is the kind of thing that a comparatively " good " German, a middle-class intellectual of old Lübeck commercial stock, was in the habit of thinking, is there any doubt of how deeply corrupted the whole springs of German intellectual life are? No light whatever is thrown upon what we are to do about it, beyond a hint that Germany should be

Germany: The Problem of Europe

reconstituted as a Federal State — a good thing so far as it goes. No doubt a crashing defeat, a catastrophe for the dominant German tradition which is responsible, is the indispensable pre-requisite; but it is, alas, only a beginning.

Any reader of Mann's book will gather what a long process of re-educating an entire nation will be necessary, and no one can doubt that the Nazis have, and will continue to have, the bulk of that nation behind them. In the best essay in the book, "This War" (and the best translated, where the rest is mediocre), Mann warns us that the German people will endure all the strain, the suffering, the privations — " their enemies may be assured that they will endure them for a very long while. They will suffer privation, shed their blood, and stand fast year after year" — in pursuit of their megalomaniac dream of power.

Mann himself has learnt one fundamental lesson: to hold his own people responsible for their acts. It is a pity that it should have needed two world wars, the shattering experience of the Nazi gangsterdom, and ten years of exile to make the conversion complete. But Germans are nothing if not obstinate.

(v)

The Institute of International Affairs has performed a great public service by making this edition of Hitler's speeches available.[1] It is perhaps symptomatic of the degradation of our times that the most distinguished Byzantine scholar in the country should have to employ the resources of his scholarship upon the outpourings of this guttersnipe of a demagogue, this criminal against civilisation who became the dominating figure in contemporary world politics. And very nobly has Professor Baynes accomplished his allotted, his sickening task. Nearly two thousand pages of lies and half-truths and near-truths, of a most cunning trickery and sleight of hand in argument, of the relentless reiteration of the *idées fixes* of an illiterate of genius, of the remorseless logic of an obsessed spirit, in this case a power

---

[1] *The Speeches of Adolf Hitler* (April 1922–August 1939). An English translation of Representative Passages. Edited by Norman H. Baynes.

The End of an Epoch

fiend; and yet this is a document of the highest importance for the politics of our time and the understanding of them. (What a comment on our time!) Even more important than *Mein Kampf* — provided that you have the political intelligence, the psychological *nous* to interpret the speeches aright. Otherwise, for the politically uneducated, quite possibly dangerous. One derives some consolation from Pitt's reflection that Godwin's *Political Justice* at three guineas a time could not do much harm. These volumes at 50 shillings, with 150 pages of bibliography, are fortunately not a book for the people. It is a dreary thought that they give us some indication of what to expect in the shape of the immense libraries that the future will accumulate around the personality of this man, as of Napoleon; whatever his end will be, there is no doubt that he has cut a figure in history! The people who under-estimated him were always fools, and did a great disservice to this country, and indeed humanity into the bargain.

It will be seen that it is impossible to review the content of these volumes in the ordinary way: that would be, alas, to write the history of Europe in the last twenty years. It will be more useful to give some general indications of their character, and the light they throw on Hitler's mind and methods. The speeches of great orators, from Demosthenes to Churchill, have usually been great literature, full of eloquence, noble sentiments, flights of fancy, flashes of humour. Not so with the speeches of Hitler: from the point of view of aesthetic tone and colour, they are like mud — grey like the field-grey of the unending battalions of German soldiers whom he has sent marching to their death. In all these two thousand pages there is only one joke, and that presumably unintentional, where he says that "To be a German is to be clear". In this waste land, this desolation, there is not a grain of humanity, not a vestige of any fineness of sentiment, of any generosity of feeling. And yet to go through it is a most powerful experience.

In what does its power reside? In the sheer power of cunning and remorseless argumentation. It is like a rat gnawing away the heartstrings of civilisation — or rather like a gimlet ceaselessly boring away, in both senses of the word

Germany: The Problem of Europe

"boring". If you are not exceedingly tough mentally, you are liable to be worn down by it; and when you hear for the 119th time that Germany was not defeated in 1918, but stabbed in the back from within, you may assent from sheer weariness and fail to note in the margin "A lie", as in duty bound. Then again, if you are not already wise as to the absurdity or even the criminality of the clichés and assumptions, or do not spot the reservations in Hitler's mind, you may be utterly subverted by the ceaseless twisting of the argument — often so continuous and so subtle that only a logician or a politician could put his finger on it. In twisting and lying Hitler is an arch-artist, and his patience inexhaustible. Who else could put across a working-class audience, for example, the lie that the Social Democrats were the enemies of the workers and that the Socialists were the catspaws of capital; while at the same time offering himself to the capitalists as the most effective way of making contact with the masses, *i.e.* breaking down their means of defence, trade unions, cooperative movements, democratic political parties? A whole anthology of flagrant lies could easily be collected, of the order of: "The German Government have assured Belgium and Holland of their readiness to recognise and guarantee these States as untouchable and neutral regions for all time". There were similar assurances all along to Poland and Czechoslovakia. Nor should this be surprising from one who declared long ago that "People will believe any lie, provided it is one big enough". The surprising, and disastrous, thing is that there were people high up in every country in Europe prepared to believe him. They acted as stooges in the process of "softening up", when they were not conscious agents.

What strikes one more forcibly in reading the speeches through as a whole is the blatant and unblushing technique of Bait. On every page in Vol. II, devoted to Foreign Policy (and one remembers Otto Strasser's saying that it was foreign policy essentially that interested Hitler), there is a bait offered to Poland, then one to Italy, next bait for Austria, bait for Great Britain. It is astonishing that our political leaders did not see through so obvious a technique. But it is combined with a more remarkable characteristic: an infallible flair for finding out the specific weakness of his opponent and appealing to that. With the Poles, their fear of Russia is played upon; with the Italians, their naïve desire to be regarded as a great Power; with the French, their deep fear of another war with its drain on French blood; with the Anglo-Saxons, their weakness for fair play, combined with the apprehensiveness of business leaders that German debts might not be paid. And always and everywhere with the propertied classes the fear of Bolshevism is exploited. It was that more than anything that made them play his game and allow him to ruin Europe — for which there will be an almighty reckoning.

As to content, there is one striking thing that is enforced upon one, especially by the early speeches, which are the most revealing: namely, that Hitler has always been a man of the Right. The whole world of his ideas from the beginning has been that of German Nationalism, brutalised and debased by the racialism of Houston Chamberlain and Rosenberg. The gods he has always worshipped are Frederick the Great and Bismarck: his mental world is fundamentally theirs, the pursuit of German power, the use of ruthless aggression, lying, fraud, trickery, violence in order to attain its expansion. Success in war is his only standard by which to evaluate human action, the only test he recognises in history. It is a nightmare view of the world. But one must never forget that its exposition in these countless speeches appealed with tremendous force to the Germans; they have supported him all the way; nor is it surprising, for what Hitler thinks and has done represents the culmination of the specific and characteristic tradition of modern Germany. Then, too, he has never had the slightest inkling of the strength of the case for Socialism: that the aim of bringing the economic resources of communities under public control is to level up general consuming power and so to diminish the conflicts of modern society and make for international cooperation. With him the whole object of gaining control of the resources of the State is not to increase the consuming power of the people but to direct those resources into increasing power for aggression, gaining ever more *lebensraum* at the expense of others, the insatiable pursuit of power. Then he has the effrontery to call this Socialism, which he equates with Nationalism. In fact there is one thing dominant in his mind — the craze for power, which answers to something deep in the German people.

If we had had something like these volumes with us before, we might have been warned. But they are not too late to be of the highest service — in prosecuting the war, in steeling our resolution to defeat such barbarism, such a crime against civilisation and humanity, and in making a peace free from illusions to safeguard the future. The work of editing is a masterpiece of which English scholarship may well be proud. The only word of criticism I have may seem a little ungracious: Professor Baynes is such a scholar and gentleman that in translating from the German he loses something of the force and vulgarity, the meanness and abusiveness of this natural denizen of the gutter and doss-house whom he is rendering: he gives him more than justice and makes his language rather less revolting and somewhat more reasonable than the original.

(VI)

How can one possibly understand the issues that confront the statesmen assembled at Potsdam in deciding the future of Germany, without knowing how Germany came to be what she is or how she got where she has? To know that, one needs to know something of Germany's history in the past hundred years, and possibly something more. And that is just what Mr. A. J. P. Taylor sets out to give us in this useful, trenchant, independent-minded book.[1]

He is one of our leading authorities on nineteenth-century foreign history, particularly that of Central Europe, and belongs to a generation of historians not educated in Germany or imbued with German ways of thought. That is the value of his book; he writes with an admirable North-Country forthrightness and gives us an English reading of German history, how the whole ghastly record of political failure and

[1] *The Course of German History*, by A. J. P. Taylor.

irresponsibility, the worship of power and conquest ending in crime, strikes an Englishman of scholarship and political sense. He says at the outset: " Not ignorance of the broad outlines of German history, but total misconception of its sense and meaning, bedevil English thought about Germany ".

The first thing to do with this book is to prescribe it as essential reading for all those who are concerned with German affairs; the second is to see that it is translated at once so that Germans may learn what their record looks like to (intelligent) people outside.

Mr. Taylor begins with a couple of summary chapters on the upshot of German history prior to 1815.

The history of the Germans is a history of extremes. It contains everything except moderation, and in the course of a thousand years the Germans have experienced everything except normality. . . . One looks in vain for a *juste milieu*, for common sense — the two qualities which have distinguished France and England.

He announces, what becomes a leading theme in his treatment, the dualism — one might almost say the duplicity — of German history:

To the west the Germans have always appeared as barbarians, but the most civilised of barbarians, eager to learn, anxious to imitate. . . . To the Slavs of the east, however, the Germans have made a very different appearance: ostensibly the defenders of civilisation, they have defended it as barbarians, employing the technical means of civilisation, but not its spirit.

We are witnessing one of the gigantic *renversements* of time at this moment, the reversal of a thousand years of history. The Slavs are re-entering upon their ancient lands.

The dominating theme of Mr. Taylor's book is, necessarily, the political abdication, the failure of one class and party after another, the aristocrats (with the sinister exception of the Junkers), the middle-class Liberals, the Catholic Centre Party, the Social Democrats, the Communists. The record of each of them is enough to make the angels weep, and none of them is spared in this plain-speaking, scarifying account of their doings.

Germany : The Problem of Europe

They all sold their souls to the devil for a price : no wonder Faust is the representative figure of German literature. The Liberals allowed themselves to be seduced by the success of Bismarck's blood-and-iron methods; the Catholic Centre gave up their federalist, democratic principles, voted for the great Navy Law which brought on the conflict with England, and voted in 1933 to suspend the constitution for Hitler.

The Social Democrats were so hag-ridden by revolutionary Mumbo Jumbo that they would not cooperate with Liberals in practical political reform; nor were they any good as revolutionaries. The Communists, instead of making democracy successful, did their best to wreck it; and when in consequence Hitler came in, " it turned out that they were old-style parliamentary talkers like all the rest ".

Perhaps Mr. Taylor does not allow sufficiently for the extreme difficulty and complexity of the German political situation, though it does transpire from his account of events. But why should the Germans at every juncture have chosen the wrong turning? Perhaps, seduced by the lust for power, the dream of conquest, they sacrificed everything for it, first freedom, then religion, law, humanity. When they did produce a statesman of first-class powers in Bismarck, it turned out in the end, as Mr. Taylor says, the greatest of disasters. Why? Because he concentrated all the powers of his genius against liberalising Germany, and in the end made it virtually impossible for the German people to achieve responsibility, maturity, self-government. Mr. Taylor sums up the consequences of it all : the emigration to America after 1848 of hundreds of thousands of liberal-minded Germans, " the best Germans, who showed their opinion of Germany by leaving it for ever "; the concentration upon administration and industry as a substitute for the political power they were incapable of winning or wielding.

They strengthened the military monarchy and urged it on to conquer others in order to console themselves for the fact that they had been themselves conquered. . . . The highest faculties of the mind, and these the Germans possessed, were put to the service of a mindless cause. . . . The disease was forced inward until it poisoned the body of Germany incurably, and the body of all Europe as well.

When one thinks of the ruin of Germany, the wreckage over all Europe, could there ever have been a more terrible *dénouement*?

Since one goes to the historian for a diagnosis of what is wrong, it may be that he has most to offer in suggesting possible ways of putting it right. " Were I a German ", Mr. Taylor says, " I should not hope for some change of heart affecting the entire nation, but should rather start afresh: go back, that is, to the municipal autonomies and small states which once made up Germany." And he speaks of the ideal of 1848 — " a free federal Germany, not worshipping power, founded on Christian civilisation and on democratic socialist principles ". Perhaps they may listen, now that they see where the worship of power has led them.

XVII

FRANCE: THE THIRD REPUBLIC

[1940]

WHAT was it that was wrong with France? It is a searching question and one of the greatest importance to find an answer to, for France herself, for us who are now left to fight alone for her future, and for the Europe that will emerge after the war. Professor Brogan's notable book goes a long way to helping us find the answer. It gives us the whole political background, over the past seventy years — the span of life of the Third Republic — to the present situation, the downfall of the Republic, the collapse of the France we have known in our lifetime.[1]

Perhaps the question should be framed rather: What was wrong with the Third Republic? For the source of France's troubles was essentially political. There was so much else that was sound and right with France: French culture, her painting and music and literature, her soil, her agriculture, the industry and thrift of her peasant cultivators, French food and wine, the art of life, which the French possess better than any other European people.

I remember having a discussion in the train, at the time of the troubles that liquidated the Front Populaire and foreshadowed the present *débâcle*, with a splendid young Frenchman, one of the administrators of Indo-China, who was coming home on leave with his wife and baby. (How good she was, too, with the baby! — so clever and competent, with the minimum of fuss.) He ended the discussion with " *Après tout*, in spite of everything, France is the best country in the world to live in ". There are a great many of all nationalities who will bear witness to the truth of that: the pity of it is that there are

[1] *The Development of Modern France*, by D. W. Brogan.

The End of an Epoch

so many hundreds of thousands of Germans bearing it out at this moment.

Mr. Brogan's book helps us to answer the political question, for it is in form a political history of France since 1870. He expressly disclaims any intention, or even the qualification, to estimate the wonderful achievement of France in other fields during this period, in art, literature, music, science. But it is salutary at this moment of defeat for France that we should remember what that achievement has been. He says, generously and rightly:

> At no time since the reign of Louis XIV has the genius of individual Frenchmen and Frenchwomen been more brilliantly displayed, or in a greater variety of fields, than this period. A history of France which finds space for the Duc de Broglie, historian and politician, but not for his grandson, the great physicist; for Calmette the journalist and not for his brother, the great pathologist; for Raymond Poincaré and not for his cousin, Henri Poincaré, the great mathematician: which has room for Zola, but not for his school-fellow Cézanne, for Senator Antonin Proust and not for his kinsman Marcel, obviously cannot pretend to give anything like a complete picture of French activity in this period.
>
> Pasteur, Debussy, Degas, Pierre Curie, Mallarmé, Bergson, the two Charcots, Alexis Carrel, André Citroën, Blériot, Père de Foucauld, Saint Theresa of Lisieux, Madame de Noailles, Sarah Bernhardt, Gaston Paris, Littré, Le Corbusier, a handful of names taken almost at random reveals the variety of talents or of genius that modern France has bred or provided a home for.

That is no more than the truth. And for all its self-imposed limits, I admire this book. Underneath its caustic wit and cynicism there is a real and singular impartiality and independence in its judgments. There is no humbug: Mr. Brogan is taken in by nobody, nor anything. At the same time there is admiration where it is due for really great men — Gambetta, Thiers, the great missionary Cardinal Lavigerie, Clemenceau, Jaurès — and for great achievements, like the French Colonial Empire, the marvellous recovery from the devastation of the last war. The book is a monumental expression of what our generation thinks of the generation before it. Immensely long — it will provide you with a winter fortnight's reading — it is interesting and lively all the time: a pleasure to read a book by a professor who knows how to write.

The Third Republic was born, as it has come to an end, in circumstances of defeat, collapse, and chaos. In one respect the situation at its beginning was even worse than the present, for it had the terrible folly and crime of the Commune of 1871 and its suppression upon its conscience. Nobody loved the Third Republic: it came into being because, as Thiers said, " it is the Republic that divides us least ".

At the beginning there were fanatical legitimists, supporters of the stupid elder branch of Bourbons, Orléanists who supported the intelligent and able younger branch, Imperialists who wanted Napoleon III's son, Republican Radicals who wanted 1848 all over again, and the Communards who never forgot and never forgave. In these circumstances the bourgeois Republic had hardly anybody's active, passionate devotion — and it has suffered from that all through its history.

Nor has it ever been a very lovable object: an unwanted child, shabbily brought into the world, shabbily endowed, run for the most part by a lot of narrow-minded and often corrupt petit-bourgeois, mean and unimpressive, with a dismal record of financial scandals and much washing of dirty linen in public.

It would have been better for France if she could have had a respectable and intelligent constitutional monarchy. It was a great pity in a way that the July Monarchy of 1830-48, under the Orléans family, came to an end, for they were intelligent and public-spirited. But their whole basis of support in the country was far too restricted, practically only the " notables " of finance and industry. And it was France's bad luck to have in the Comte de Chambord (Henri V), the heir of the Bourbons, a man who was an utter fool politically, completely cut off by his character of mind and education from the realities of the modern world, and the sane and sound currents of public feeling.

Nor, it must be confessed, were the French aristocracy, his supporters, much better. What strikes one in reading this record of their behaviour is their criminal levity and folly. Because they were defeated, they were prepared to go to any lengths to see the Republic defeated — even selling out to the national enemy! One can see now how this strain has persisted among them and where it has led.

There was that leader of Parisian society, the Duchesse d'Uzès, whose great wealth supported the mock-Hitler of the 'eighties, Boulanger, a sort of Mosley of the time. Then there was the undying determination of the French upper classes, which kept France in turmoil for a decade, that Dreyfus should not have justice. There is the record of inconceivable scurrility of the Royalist paper, the *Action Française*: and of its editor, Charles Maurras, ending up with his open invitation to assassinate Léon Blum, the distinguished and upright leader of the French Socialists: " Any kitchen-knife would do ". Two years ago Maurras was elected to the French Academy.

What has been fundamentally wrong with the Third Republic all through — Professor Brogan does not say it, but the conclusion arises inescapably from his whole book — is that sectionalism, party feeling, class resentment have gone so far as to undermine the foundations of the State itself. Elements in the French upper classes would rather see the enemy win than be ruled by the Left; the idiot Communists were so irreconcilable as to aid the triumph of their worst foes, the Nazis and Fascists (Brogan has a good phrase for their foolery, " as the Comintern, having launched a disastrously ineffective campaign in one country, turned to another to try out its well-tested recipes for defeat "). The Radicals would not cooperate with the Socialists; the Socialists would not join forces with the Radicals.

What idiots they all were, a rational person must think — and what sufferings people bring upon themselves by their (political) folly. The English aristocracy have not got such a bad record — they were prepared to make concessions and to cooperate. But then, they had not been defeated as the French aristocracy were; besides, Ulster was not a pretty show, nor the behaviour of the English ruling classes since 1931 anything to be proud of. The simple truth about high politics, as with marriage, is that for the show to be a success there must be much give and take. And that French politics has never had. Behold the consequences!

There are other reasons, too, in the nature of the French political system, about which Professor Brogan is explicit and even damning. In some ways the French parliamentary system was a caricature of a Parliament, and merited the strictures passed by Hitler upon Parliamentarism in *Mein Kampf*.

From the very beginning to the end the French deputies kept as much power as they could in their own miserable, and often well-greased, hands (hands " calloused with wire-pulling ", as McTaggart said of Ramsay MacDonald). They were jealous of any really big men coming into power: Gambetta's great ministry was never formed, his plan of a unified party of the Left, like an English party, never came into being; the deputies turned their backs on Clemenceau the moment after he had saved the country (" Did I hear a cock crow? " said that mordant old tongue to some deputies who protested their regret). Jaurès was never in office, and he was perhaps the greatest of them all: the one figure whom Western Socialism has to place against Lenin, a more genial and, when all is said, a more hopeful figure than the great Russian.

All this meant brief spells of government, short ministries in which little could be executed, nothing planned, responsibilities shirked — until the abyss was revealed in a lightning-flash three years ago at the time Hitler invaded Austria, the first of his foreign conquests, when France was found without a government because no one would take the responsibility of what was about to happen. That was a pointer to the end.

Mr. Brogan tells us that the root of all the financial scandals that have so troubled the Republic, from Panama to Stavisky, was the same: " the horror of responsibility; ministries came and went; authorisations were given and refused "; people lost track of what was happening, the shady speculators, the blackmailers, the swindlers flourished. The horror of State-power was the bedrock of Radical political doctrine. To that extent the Radicals, the largest of French parties, bear the responsibility for the downfall of what was above all *their* Republic. The Socialists were right about them. And since, as Brogan tells us, it was the Radicals who were "the best judge of what the average Frenchman will stand for", the average Frenchman has his responsibility too. All that suffered from the clever, senseless game of parliamentary manœuvring was, he says, "the independence of the executive, the strength of the administrative system and the long-term interests of France".

The book might have had for its motto that revealing sentence of Flaubert describing the attitude of a meeting in 1848 towards a distinguished professor who was addressing it: "*Bien qu'on l'aimât tout à l'heure, on le haïssait maintenant, car il représentait l'Autorité*". How like the French that is! In a way it is the moral of the whole history of the Third Republic and accounts for the weakness in so much of French political behaviour.

What is necessary if things are to be put right is a stronger Executive. "Only an Executive far stronger than that provided by the French Parliamentary system could protect the authority of the State and the integrity of the administration of justice against the recurrence of scandal of the same kind", Brogan says apropos of the Stavisky case. "The diffusion of authority in the French Chamber bred crooks, as the concentration of it in the House of Commons breeds yes-men." A shrewd point that from the Professor of Political Science!

It has been evident to those who have watched France in the post-war period that there was something profoundly wrong, and of a political and social character. Professor Brogan brings out, and it is right to remember, the terrible blood-drain of the last war upon her vitality: proportionately more than twice that upon this country, and we are in a position now to know what we lost in leadership and vigorous conduct of our affairs in consequence. France made a very wonderful effort towards recovery after 1918, for which she has not received due credit. After all, France is always more worthy of respect than Germany: after 1870 the French did get down to reconstructing the life of the nation upon solider foundations than before; they made their greatest achievements in painting, in music, science, and poetry, after defeat. The Germans after 1919 sulked and could think of nothing but their vanished vision of a Europe in which they could hector and bully smaller peoples as they pleased; they never acknowledged even to themselves their responsibility for the War of 1914–18. The French were much more prepared to cut their coat according to their cloth; they have rather fallen over backward in their readiness to do so in the last few years.

France's renewal, we may be sure, will not come from the Old Men of Vichy, who are the débris and in some cases the worst culprits of the old régime. It will come from new and unknown men, who are not inculpated by the mistakes of the past and who will have a fresh page to write history upon. What we can be sure of is that the France that will emerge will not be the France of the Third Republic, and we hope that they will not make the mistakes of that half-and-half régime. Underneath the surface of her past political systems, absolute and constitutional monarchies, first and second empires, republics of which we now witness the end of the third, *Il y avait la France*, as a famous phrase has it. Professor Brogan is profoundly right when he says of that eternal France, " the nation to which our Western civilisation owes most ".

There is only one correction to be made in his Preface. Owing to the collapse, instead of France being the main sword and shield of that civilisation, it is now we ourselves. We have reason to be proud that the honour of defending the magnificent heritage of French civilisation with our own has fallen to us.

XVIII

THE LITERATURE OF COMMUNISM: ITS THEORY

[1929]

It is impossible to make a survey of Communist theory compact and homogeneous like Mr. Eliot's survey of Fascism,[1] for compactness and homogeneity are not in the nature of the literature of Communism. This after all is only to be expected: there is such a difference between these two in the length of time that they have been under consideration, in the amount and importance of their theoretical production, and it may be added, in the universal significance of the one and the fairly limited relevance of the other. Fascism is strictly a phenomenon of the post-war world; and one may gather a good deal of its meaning by remarking that it is confined to those Western countries that one would have expected, on other grounds, to be not particularly capable of running parliamentary institutions. But Communism has been in the world as a political programme since 1848; in that form it is a product of the class divisions within modern industrialism, and so it has a significance, both practical and theoretical, that extends wherever the conditions of capitalist industry are to be found. Further, in that course of time, it has produced a prodigious literature of theory and polemics, which it would need a whole book to survey. But here I am discussing only the theoretical aspects of Communism, and estimating its possible impact upon English thought. In short, what is its significance for us: and what use can be made of it?

Any student of Marxism, to take Communism on its more academic side, cannot but be struck by the extraordinary way in which English thought has escaped the influence of Marx.

[1] v. *The Criterion*, 1929.

The Literature of Communism: Its Theory

It is not unaccountable; but it remains remarkable that during a half-century, when in all the chief European countries the work of Marx and the controversies around his school have been one of the most potent intellectual forces of the time, in England its influence has been a mere trickle. Outside of Hyndman, and certain strands of the Fabian web, it can hardly be said to have existed. And this, in spite of the fact that Marx lived here for more than thirty years of his life, based a good deal of his work on English economic theory and industrial experience, and in the end attained to a certain respect for the country of his adoption as against that of his birth.[1] It may be that with us, the greater and not dissimilar influence of Darwin absorbed attention to the detriment of Marx. But the main reasons for his neglect are probably more general. There is in English minds a notable reluctance to systematise our reflections upon industry and politics; and when the system takes the highly abstract and semi-philosophical form of Marx's theories, we are apt to consider the whole thing as unpractical and irrelevant. Further, we have a temperamental dislike for the realist and the analytical; especially if the realism and the analysis are applied to the assumptions of our political system. The natural business man's outlook of late nineteenth-century England was empirical. And when it trespassed from business into academic spheres, it was inclined to be idealistic and religious in its ways of thought — also an expression of the business man's mentality. There is a kind of idealism that is natural enough to a community on the up-grade of economic and political power; just as there is a form of realism, sometimes appearing as a systematic cynicism, that is appropriate to a nation that feels itself to be on the down-grade of power. It is precisely because the situation of England in the world today is something between these two positions, neither on the up-grade nor the down-grade, but upon a difficult and dangerous equilibrium, that the Marxist analysis of politics and economics in terms of interests, and conflicts of interest, is particularly relevant and may be helpful. And it is for the same reason, a general historical

[1] Cf. *Capital*, p. 863. Translated by E. and C. Paul.

consideration, that Marxist influence has been so negligible up to recent years: the progress in our social conditions and the advance of our wealth and power up to the immediate pre-war period, prevented the contradictions within the system and the conflicts of interest among classes from appearing at all clearly. The relevance of Marxist theory was therefore not generally obvious; and its use and application were confined to small circles which came into close contact with industrial conflicts, or to those who understood that more general conflicts so far from being non-existent, were merely latent.[1]

It is the great strength, and at the same time the weakness, of Communism, that it was born out of division and struggle. The *Manifesto* of the party, which is its real starting-point in international politics, was drawn up in January 1848, at the very beginning of the year of revolutions. And so it has always been at home in periods of unrest, maladjustment, and crisis; and its theory explains the phenomena of social discontent in a way that is unknown to most social theory, and is only beginning to be comprehended by average commonplace social thought. One cannot but feel this with great force on reading the first section, say, of the *Communist Manifesto*. It is one of the great classics of political thought, most of which have been similarly short concentrated pamphlets. But if one compares this with any one of the others, take the *Contrat Social* for example, one is impressed by the exact relevance of the one and the comparatively academic character of the other. It is not that one underrates Rousseau's political theory; indeed it may well be that one agrees with his conclusions rather than with Marx's. But before one can get at the core of the *Contrat Social*, it needs a considerable amount of reinterpretation, and of transposing into another key before one can grasp what significance it has. Whereas with the *Manifesto* one sees immediately how significant is its whole approach to politics: it is the first such classic to be based upon a diagnosis of modern industrial society. It reveals a mentality that is equal to

---

[1] Cf. *Communist Manifesto*, Marx and Engels: " in particular, a portion of the bourgeois ideologists, who have raised themselves to the level of comprehending theoretically the historical movement as a whole ".

The Literature of Communism: Its Theory

industrialism — naturally enough, for it is in a sense the only political theory, along with Socialism, that has taken industrialism into account.

The pamphlet begins with a historical summary of great acumen, based upon the materialist conception of history. The history of society is the record of class struggles, is its first proposition: "Man is born free but is everywhere in chains". But whereas the class conflicts of previous ages have been manifold and complicated by hierarchical arrangements of society, this epoch is distinguished by the simplification of class antagonism into that of the bourgeoisie and the proletariat.[1] "Each step in the development of the bourgeoisie was accompanied by a corresponding political advance of that class. An oppressed class under the sway of the feudal nobility, an armed and self-governing association in the medieval commune; here independent urban republics (as in Italy and Germany), the taxable 'third estate' of the monarchy (as in France); afterwards in the period of manufacture proper, serving either the semi-feudal or the absolute monarchy as a counterpoise against the nobility, and in fact, corner-stone of the great monarchies in general, the bourgeoisie has at last, since the establishment of modern industry and of the world-market, conquered for itself in the modern representative state, exclusive political sway. The executive of the modern state is but a committee for managing the common affairs of the whole bourgeoisie."

In this paragraph, as in many another one might have chosen, the typical qualities and defects of the whole work are evident. In one sentence is concentrated the whole history of the modern capitalist class; in another is expressed a whole theory of the State. Not bad going; one is not so much surprised that the theory of the State is only half the truth, as that the historical generalisation is fairly recognisable. The chief ground of criticism against the *Manifesto* is on the score of

---

[1] "By bourgeoisie is meant the class of modern capitalists, owners of the means of social production and employers of wage-labour. By proletariat, the class of modern wage-labourers, who having no means of production of their own, are reduced to selling their labour-power in order to live."

over-simplification; but this is inevitable, a manifesto being what it is and not a treatise.

Next, the revolutionary character of the bourgeoisie in the past is emphasised: an important point we are apt to forget. Who would think, to hear a Conservative defending the rights of property, that the present régime of property is founded upon a series of revolutionary acts, the expropriation of the Church, of royalists and of common rights? And yet so it is; we tend to ignore how revolutionary the history of England has been even while reading it, until Marx reminds us. There follows a paragraph or two on the revolutionary character of the bourgeoisie in the present: how "it cannot exist without constantly revolutionising the instruments of production, and thereby the relations of production, and with them the whole relations of society". There is a magnificent admission of the productive activity of capitalism: "the bourgeoisie, during its rule of scarce one hundred years, has created more massive and more colossal productive forces than have all preceding generations together". But it is owing to the ceaseless and uncontrolled extension of such productive forces that capitalism comes upon its nemesis: its lack of controls produces constant crises, local and general, and on the other hand, the conditions of modern industry have created a proletariat that is recruited not only from the working class, but also through the depression of the lesser middle classes into it. In proportion as capitalism extends its hold upon the exploited masses, they tend gradually to unite into a politically conscious class. They must be organised as such; and as this becomes a movement of the immense majority, whereas "all previous historical movements were movements of minorities", it prepares the "overturn of the bourgeoisie and lays the foundation for the sway of the proletariat".

To summarise a summary is a forlorn and ungrateful task; but the above may pass as an account of the first and most important section of the *Manifesto*. The rest has rather a disproportionate amount of polemic and of sectarianism, of which the literature of Communism is always too full.

What is so important about this political manifesto, apart

The Literature of Communism : Its Theory

from its historical and propagandist importance, is its method of approach to politics, its analysis of society. No vague idealistic manual of politics, tricked out with the glamour of historical tradition now in fashion, throws any light in these days of confusion upon the darkest and most critical problems of modern society, those of the relations between classes. We all hear enough, and more than is good for us, of the national good and of the natural conservatism of the English people. It would be more use to get down to the work of analysing what interests we mean when we say " the English people " ; and whose interest would be benefited primarily by a given course of policy. For it is quite clear that certain interests, certain groups as defined by economic function and historical tradition, have only to be isolated out of the mass of interests that is the community, for it to become apparent that they are either non-productive or anti-social. And towards an analysis of this sort as a basis for politics Marx showed the way. Naturally enough, in its beginning it was too crude ; and the rough-and-ready way in which the *Communist Manifesto* treats class divisions is open to criticism. That, however, does not destroy its value; one can agree with a good deal of its diagnosis without accepting its remedies.

My chief criticism of the position is that it is far too simplified. One can agree with Marx in his analysis in terms of class and class interest, without accepting the conclusions of the Communists that there is one single class division in society that must be accentuated to the point of revolution, in order to achieve its forcible abolition. For, first, there is no one single class conflict in society, but several ; of which some are more important than others. And second, the forcible abolition of class by means of dictatorship is not the only way of bringing class division to an end. That has, for a variety of reasons, been the way in Russia. But fortunately there is another way : on the basis of this analysis, one may construct a reasoned politics which makes possible the gradual elimination of class division from society, by means of a system of economic and political controls such as Socialism offers. That is the more orderly and rational method : it happens to be necessary

The End of an Epoch

for the welfare of Western society, and at the same time, owing to an effective political tradition (in England at least), to be possible.

The first point is more difficult; but it is clear that there are contrary pulls of interest in society that make the system a more complicated pattern than that pictured in Marxist theory. There is a sense, for instance, in which capitalists and workers in one industry may have a common interest as against capitalists and workers in another; while preserving at the same time in each group the division between capitalists and workers as to the share of the product. That is, a twofold division of interests; the coal-mining industry is a case in point which comes most easily to mind. Further, is it so certain that in every sense the interests of the workers are at one? In a certain passage of *Capital*,[1] Marx took it rather lightly for granted that skilled labour was reducible to " average unskilled social labour"; and though this may be so for the purpose of his argument, in that skilled labour would incorporate higher costs of training, it does not help with the essential question how far there is a divergence between the interests of skilled and unskilled labour. Lastly, the *Manifesto* admits that " Wage-Labour rests exclusively on competition between the labourers ". And it really seems in these latter days of capitalism as if competition between labourers is being limited; in which case, the necessity for " revolutionary combination " does not exist. On the whole it seems that Trade Unions have been fairly effective in stabilising rates of wages over the field of their organisation. Unemployment insurance, too, in the post-war period is to some extent a safeguard against undermining standards of living. Most important, however, is the probable effect of a stationary population, and perhaps even a declining one, upon the competition of labourers with each other. It may be almost entirely to destroy their competition, and correspondingly to advance their standards and claims. It would need only just such a revival of trade as one can without undue optimism envisage, plus the raising of the school-leaving age to fifteen, to reduce the competitive surplus

[1] P. 193.

of workers to manageable proportions. There may even, in fifteen or twenty years, be a need for more workers than we have at our disposal. It is this consideration that makes the arrival of the Labour Movement to power almost a certainty, rather than that of a revolutionary proletariat.

But for the full development of the academic side of Marxism, we must go at length to *Capital*. Happily within the last few months a new translation [1] of the first and most important volume has appeared, which lightens considerably the labour of understanding the system.

Perhaps the best introduction, for a diffident and reluctant reader, to the charms of *Capital* is by way of the footnotes. These are extraordinarily rich and varied in their character; they draw upon inexhaustible stores of information; and very often they are much more revealing than the text, of the assumptions underlying the argument and the ends Marx has in mind. One comes upon a penetrating suggestion for the study of law as the product of the material relations of production; or a reply to the common misconception that historical materialism holds good of modern industrial society, but not of previous ages; or an amusing note on the preaching of Malthusianism by parsons of the Church of England, as contrasted with their own prolific practice; or quotations from *Timon of Athens* and the *Antigone* to illustrate various conceptions of the nature and use of money. Everywhere there is the evidence of universal reading and reflection: there is nothing that is not grist to the mill.

If one asks what is the purpose that Marx had in view in writing *Capital*, it is clear that he was attempting to analyse the economic processes of society. He refers, for instance, to " our analysis of the basic form of capital, our analysis of the form in which it determines the economic organisation of society ". In his first preface, of the year 1867, he says specifically, " The final purpose of my book is to reveal the economic law of motion of modern society ". That is, the character of the work is in the first instance sociological; the

[1] Karl Marx, *Capital*. Vol. I. New Trans. by Eden and Cedar Paul. Allen & Unwin, 1928. My references are to this edition.

three volumes of *Capital* proper deal with different aspects of " Capitalist Production ", and the work is only in the second instance a " Critique of Political Economy ". And so those people who approach the work expecting to find a complete system of pure economic theory, are not unnaturally disappointed, and find the book faulty for not being what it does not pretend to be.

It is, of course, very largely economic theory, but in this specialised sense; it is, also, very faulty: but what major work of this kind in the nineteenth century is there that is not? Constantly one finds in reading *Capital* that the theoretical conclusions of orthodox economics stand, where the theories of Marx fail. But this does not destroy the value of the work; the *Political Economy* of Mill, for all its mistakes, still has more than antiquarian value. And there is a great deal more in Marx that is left untouched by the ebb and flow of pure theory.

The historical method is often held out as the typical achievement of the nineteenth century in thought;[1] and no one, not even Maine, contributed more to its formulation or applied it on a grander scale than Marx. It is partly on account of the historicity of his mind that he is so impatient of bourgeois economics which regarded the régime of early capitalism as the final end of social life. With regard to its theories, he says: "It is writ large on the face of these formulas that they belong to a type of social organisation in which the process of production is the master of mankind, and in which mankind has not yet mastered the process of production. To the bourgeois mind, however, they seem as self-evident as, and no less a natural necessity than, productive labour itself. That is *why* bourgeois economists treat pre-bourgeois forms of the social productive organism much as the Fathers of the Church have always treated pre-Christian religions." More profoundly than this, he examines the characteristic relation of capitalist to worker in modern society, the owner of money or commodities as against the worker with only his labour power to sell, from the historical

[1] Cf. Morley, *Notes on History and Politics*; and Vinogradoff, *Historical Jurisprudence*, Introduction.

point of view: " The relation has not a natural basis, nor is it one met with in all historical epochs. It is manifestly the outcome of an antecedent historical evolution, the product of numerous economic transformations, the upshot of the decay of a whole series of older forms of social production."

There are two consequences of this view, besides its effect upon the analysis of current capitalist production : it involves the study of pre-existing forms of production, and the construction of possible future forms for it to take. That is, roughly, the historical side proper of his work, and the propagandist or Communist side. With regard to the first, his influence has been enormous; the history of capitalism was, so to speak, his discovery. Such a passage as that which opens Chapter IV has been the starting-point for whole libraries of books, and for the acquisition of the stores of knowledge which we now possess on nearly all historical developments of capitalism. Most of the great names in economic history, Weber and Sombart, Sée, Ashley, Unwin, and Tawney, are indebted to Marx.

Yet a notable result of this historical conception of capitalist society, paradoxically when one considers the large dogmatism of the book, is in a healthy scepticism. There is no longer any case for thinking of capitalism as the *terminus ad quem* of social movement in the nineteenth century — or for that matter, in the twentieth. The capitalists of the period were too busy extending the scope of production, and their allies the economists too closely engaged in casting up the accounts, to look before or behind. And Marx's historical scepticism has been justified by the event: we in our time can see the changes. It is not so important therefore that what Marx thought would take place, should not exactly have been brought about.

This question would appear to belong to the propagandist, or rather, the prophetic side of Marx's genius. But even this side of his activity is based upon the description and analysis of economic phenomena. *Capital* begins with an analysis of the nature of the commodity, in which Marx finds " antagonisms " in its very being; that is, divergent claims of the various participators in production upon the product.

From this he builds up the structure of conflicts and cooperations which forms society. The Labour theory of value, which he takes over from Ricardo, has very little to do with this; and so its inadequacy in theory does not invalidate the economic analysis of society which is the real aim of *Capital*.

He carries this forward to two further developments. There is the point that "the product is the property of the capitalist, not that of the worker who functions as direct producer". Marx apparently did not envisage a state of affairs in industry in which the workers should share in the product with the capitalist. And yet, in some developments in industry [1] there is considerable likelihood that this may be advanced from the side of the capitalists. Socialists display less interest in this possible line of development, for they look more to public control of national industries. But the effect of its application upon the completeness of the Communist case against capitalist industry is not to be neglected. The second development is connected with the complex of questions around surplus value. Marx's account of the way in which value is made up is a consciously abstract one, dependent upon certain definite assumptions such as that of the "socially necessary", which seems to imply an ideal of the most economical working of institutions as they are. Obviously the account has a logical rather than an empirical importance: it is intended to establish symbolically, rather than realistically, certain points like that of a surplus element whose presence in some form in production it is difficult to deny. The whole argument is an interesting attempt to carry further the Ricardian analysis on these lines. If it has been superseded by another and more successful approach, it does not mean that there are not very important contributions made by the way. And further, whenever the fruitfulness of the post-Marshallian approach to economics should come to an end, there may be wisdom in going back to the previous method and attempting a synthesis of the two. It may be that there is a way of advance for economics, as there certainly is for politics and history, in

---

[1] Cf. *Britain's Industrial Future*; and cf. the theory and practice of Lord Melchett.

bringing into juxtaposition and correlation the best elements of the academic and Marxist schools.

But there is more in Marxism, and in *Capital* as a book, than so much intellectual virtue. Otherwise they would not have the appeal of a creed and of a bible which their enemies are careful to allow them. The contrast which is ever present to the mind of Marx, as with any other social reformer, is that of the bondage that is with that of the freedom that might be. He sees this antinomy on the plane of the economic and social struggles of men : " Things cannot be otherwise in a method of production wherein the worker exists to promote the expansion of existent values, as contrasted with a method of production wherein material wealth exists to promote the developmental needs of the worker. Just as, in the sphere of religion, man is dominated by the creature of his own brain ; so in the sphere of capitalist production, he is dominated by the creature of his own hand."

One cannot grasp the ultimate meaning of Communism apart from what one may call the values of a positive humanism. And yet the denial of values plays a great part in Marxist polemic and propaganda. The truth seems to be that this denial is itself polemical, or purely tactical. They are engaged in driving home the facts of a class analysis of society ; they wish to concentrate attention on the extent to which interest rules in social life. And so the well-meant idealistic appeals of Christian Socialism only meet with their scorn and hostility. In a sense, the Communists are right; for such idealism serves to befog the issues, when the essential thing is that the people should understand the realities of a capitalist world. The fact of exploitation once clearly understood would go a long way towards making exploitation impossible. Certainly the Communists themselves, as individuals, are guided by ethical motives. One cannot read Clara Zetkin's account of the family life of Lenin [1] without feeling that he for one was impelled by the desire, among other motives, to do good to the Russian people, to see them taught, fed, and freed. But again the Communists are right in pointing out that it is

[1] Clara Zetkin, *Souvenirs sur Lénine* (Librarie Stock).

The End of an Epoch

not this that makes a revolution; there must be the necessary materialist pre-conditions.

These two elements may be too rigidly divorced in their theory. For within a given group interest, for instance, there is room for loyalties, jealousy, sacrifice. And certain it is that the Revolution in Russia, like most revolutions, has set free tremendous psychological or spiritual forces in the world. Whenever one reads a book of any value on the conditions there,[1] one notices the signs of a great optimism among the people of the Revolution. Perhaps it is only natural; but they are a great and gifted nation, and one cannot but think that a race of a hundred and thirty millions will be very important in world politics of the future. So it would appear equally important that we should understand the theory by which this revolutionary régime is guided. As far as England is concerned, it looks as if we have turned the corner where Communism might have been dangerous; even those " indigenous " schools of political thought, to which Mr. Eliot would like to see an " alternative ", may have surmounted the difficulties. And besides the practical importance of Communism in Russia and as an international party, it is always likely to exert an intellectual influence in some of the ways which I have attempted to suggest. So that of Mr. Eliot's conclusion that " Russian Communism and Italian Fascism seem to have died as political ideas, in becoming political facts ", it is only with one-half that I agree.

[1] Maurice Dobb, *The Economic Development of Russia since the Revolution*; Ralph Fox, *The People of the Steppes*, and *Storming Heaven*, a novel; Alexander Wicksteed, *Life under the Soviets*; Maxim Gorki, *Écrits de Révolution*.

XIX
THE THEORY AND PRACTICE OF COMMUNISM
[1930]

THE prospects of Communism in the modern world depend upon events in Russia; and the Russian experiment has not yet emerged from crisis. Indeed, with the initiation of the Five Years' Plan for the industrialisation of the country, the Revolution has entered upon a new crisis, and perhaps a final and decisive one. Up to 1925 there was some possibility of accommodation with the Western world. Internally, the Revolution was passing into a Thermidorian phase; and the policy of Stalin at that time was to press forward with the New Economic Policy of Lenin, allowing private trading and relying upon the accumulation of private capital to build up the resources of the country. By following a moderate course he hoped to attract foreign capital on a large scale, by which alone it was thought the economic reconstruction of Russia could be achieved. This line of action was never conceived as a retreat to capitalism; it was intended as Lenin expressed it in 1922, *pour mieux sauter*. But it gave rise to much heart-searching among the old guard of revolutionaries, and was the starting-point of the Left Wing Opposition led by Trotsky and Zinoviev, which conducted a struggle against Stalin and the Government right up to 1927, when they were defeated and exiled. But in 1924, with a Labour Government in office in England, it looked as if the policy of accommodation would succeed; negotiations for a credit and with regard to a debt settlement were set on foot that might have brought a liberal solution to the Russian problem. The Labour defeat at the election and subsequent Conservative policy made this impossible; and once again Western capitalism prevented a compromise solution and re-established the financial blockade.

Mr. Michael Farbman makes this the main consideration in the reversal of Bolshevik policy from 1927 onwards. But it is hardly likely that external events like these were all in all, though they may have been the marginal factor, so to say, in bringing about the new orientation. One must look for the mainspring of events in the internal evolution of the country. The strength of the Opposition lay in its appeal to the proletarian element in the Revolution; its weakness was its reliance upon world revolution at a time when the wave of world revolution was subsiding, and in default of which it was without a practical policy. Owing to the realism of his outlook and his control of the party machine, Stalin won out; but in the circumstances of 1927 and the final extinction, with the Arcos raid, of any hope of compromise with capitalism, he won at the price of adopting a large part of the Opposition's policy. The Five Years' Plan which was then brought forward, represents a reversion to collectivist ideals: it is the return to Communism. Its motive is to build up capital from Russian resources alone; since the Bolsheviks cannot get any from outside, they have had to begin at zero, accumulating little by little the surpluses of their industry and agriculture, so as to finance industrial development and make themselves self-sufficient. It was not until 1927 that general production reached the pre-war level; and they have made that the basis for their advance towards industrialisation. The process is, quite simply, to take it out of the workers, and still more the peasants; and the political problem is to know how far the Government can go in squeezing them, without producing a reaction. So far, they have been able to go further than even they expected; and the returns on the first year's working of the Plan, it is generally agreed, reached 40 per cent more than their estimate. The whole thing obviously involves a terrific strain to the economic system; for the peasants may respond by not producing, and then collectivist farming has to be resorted to simultaneously with industrial collectivisation. Then again the peasants may endanger the collective farms, so one gathers, by flocking into them in greater numbers than the Government

The Theory and Practice of Communism

can deal with. It has become a war of attrition; and Stalin has announced that they can only win if they have nerves of iron. Whatever it is, it is certainly Communism now. The consequences in themselves would be terrible if they should fail; on the other hand, they may succeed.

But it would not be at all pleasant to contemplate if they should succeed, either. It would mean the stabilisation of a great community, self-sufficient in mind and power, and cut off from Europe by its remorseless struggle. The generations would have grown up in the atmosphere of war made permanent; with an active, working hostility to the outside world, and for the first time with the ability to express it in force. It is no easy outlook for the world.

This difference in character has emerged between the policy advocated by Trotsky, and as it has developed in Stalin's hands. Both were based on a reversion to Communism and the working-out of the proletarian revolution; but whereas the former was idealist and international, the latter is realist and Russian. Throughout the struggle around the Opposition, Trotsky remained the man of 1917, a figure of the heroic age of the Revolution. But there would be little political significance in the struggle if it were only a duel of personalities; the point is that in these two figures, two different phases of the Revolution find expression. It is to be remarked that what the Revolution needs in its present phase, is not the brilliant orator of the early days, the insurrectionary leader, *l'organisateur de la victoire*, but the trained administrator, the man who can supervise the whole machinery of production and distribution. Hence the importance of Stalin. And in comparison with that, the superb dialectical case that Trotsky makes out for himself, and all the discussions as to which is the real inheritor of Lenin's policy,[1] matter but little. It is clear that Trotsky was in fact more closely associated with Lenin; and it is said that Lenin at the time of his last illness was preparing an attack against Stalin, because of his dangerous control of the bureaucracy. On the other hand, it seems probable to an impartial observer that Stalin adheres more

[1] *v.* L. Trotsky, *La Révolution défigurée*; Joseph Stalin, *Leninism*.

closely to the conceptions of Lenin on such test points as the dictatorship of the proletariat, and the question of world revolution. The real foundation of Stalin's power is not his theoretical rightness, but that he answers to present needs; and it is only good Marxism to point that out. This does not imply that Trotsky's role is played out, for if the Five Years' Plan should come to disaster, or there should be a new wave of world revolution, there would be his opportunity and his star might be in the ascendant again. Meanwhile Stalin is in power; a nationalism of the Revolution has developed; world revolution has subsided.

Well, has it? It is precisely on that point that the schools of Communist thought differentiate. Trotsky has never been willing to admit the danger in relying upon the revolutionary movement in other countries: what is the point of a revolution in one country alone, to Communists who want the end of capitalism, he asks. He has expressed his view of the essential relation of the October Revolution to world movements in his " theory of permanent revolution ".[1] And in action, too, Trotsky has always been more intransigent: one has only to look up his attitude towards England in *Where is Britain Going?*, or to the General Strike in 1926. Or one may compare his view of the " revolutionary situation " in Germany in 1923, and his rebukes to the German Communists, with the more realist position of Stalin. The latter, in turn, fortifies himself with a theory and with a long quotation from the *Collected Works* of Lenin.[2] But Stalin's theory is a more workable one, and may in the event have some importance for the future relations of English and Russian Socialism: it is the " theory of socialism in one State alone ". And indeed, with or without Lenin's authority, it is a strong position, for if Socialism is not possible for one country unless all the others are socialist also, what is the future for Russia? A dilemma as to policy which Stalin does not fail to put, for he certainly has a better answer than Trotsky has. Moreover, its bearing upon the case for English Socialism is obvious; for if Socialism may be achieved

---

[1] Trotsky, *op. cit.* p. 17 and pp. 100-101; Stalin, *op. cit.* pp. 184-98.
[2] Stalin, *op. cit.* p. 183.

in Russia alone, it may be in England, and if this is compatible with Marxist orthodoxy, then it should diminish Communist propaganda against English Socialism. In spite of the Five Years' Plan, there should be greater possibility of accommodation between England and Russia with Stalin than with Trotsky.

One may extend the question *how far world revolution has subsided?* into the questions: *can capitalism stabilise itself?* and if it cannot, *how and by what will it be superseded?* The Communists diverge over the first question; but it is according to the answer to the last two that Socialist thought generally differentiates itself.

This is the fundamental question, according to Trotsky; and all Communists would agree with his statement of it in the preface to *La Révolution défigurée*: [1]

L'essentiel est de savoir si le capitalisme est capable de sortir l'Europe de l'impasse historique. Si les Indes sont capables de s'affranchir de l'esclavage et de la misère sans déborder les cadres d'un progrès capitaliste pacifique. Si la Chine est capable d'atteindre le niveau de culture de l'Amérique et de l'Europe sans révolution et sans guerres. Si les États-Unis sont capables de venir à bout de leurs propres forces productives sans ébranler l'Europe et sans préparer une effroyable catastrophe guerrière à l'humanité tout entière. . . . L'essentiel est de savoir si, en tant que système mondial, le capitalisme est encore progressif?

However, these questions are rhetorical, for all Communists would agree that capitalism is incapable of stabilising itself for any time. And for this view they would rely on orthodox Marxist theory that the contradictions of capitalism are so fundamental that they are bound to break it; that the conflicts revealed by the development of economic imperialism, the last stage of capitalism, are irreconcilable: for example, the rivalry between one group and another, say Great Britain and the United States, as also the conflict within an imperialist system, say between Great Britain and India.

It must be granted that they have a strong case, even if one is tempted to observe that at the same time Communists

[1] P. 18.

point out the dangers so forcefully, they seem not at all anxious for the world to avoid them, but rather to invite them upon us. Still, Trotsky's questions search to the heart of the matter: this is a world of economic conflict. There is the dangerous challenge of the United States to English export markets; and we live by our export trade. There is the intolerable economic impact of America upon Europe, and of Western Europe upon Germany. There are the conflicts that run like a fringe of flame all round the outer margins of empire: China, India, Kenya, South Africa, Haiti, the Philippines. The Communists will be right if the world does not learn how to avoid war. And if it cannot, and there should be another world war, we may all become Communists. But that alternative would be not much less disastrous: it would be a counsel of despair, for it too would involve the acceptance of force. Now I am not denying the importance of force in the historical process: my position is not that of an *a priori* pacifism. I should merely have thought that it was the very definition of progress (and also good Marxism) to work out ways of canalising economic conflict into political channels, and on that basis to seek agreement. Whether this is possible depends on whether the conflicts are avoidable.

The Communist view that they are inevitable is based upon a too rigid conception of the nature of capitalism; it comes from interpreting Marx like the word of God, and I am all in favour of the Higher Criticism. Capitalism is no more static than Communism; it goes on developing, providing itself with better improved means of canalising conflicts; for example, there is the model English imperialism in West Africa, which aims at training the natives to self-government, to place against the crude methods of East and South Africa. Or there is the liberal politics of withdrawal exemplified in Ireland, Egypt, and China, and that is the declared end of British policy in India. So much for the conflicts *within* an imperialism. As for the struggle of one empire with another, it may be that one will become so strong that any struggle against it will be impossible. This is the typically defeatist view of Bertrand Russell, or was, in *Prospects of Industrial Civilisation*. Its defence is that we are too weary of war to continue the struggle, that a defeatist mentality anyhow exists, and can appeal to the argument, not unakin to Marx, that more important than to live well is to live at all. This is not the spirit of which great deeds are made; yet in certain circumstances it may be the common sense that will save the world. It is poles apart from the idealist mentality of Communists, who would argue that you should be prepared to sacrifice yourself and everybody else in order that the cause which is greater may live, and later generations benefit. This is demanding a lot for the sake of an hypothesis! A Communist England, for instance, would be in danger of having its food supplies cut off; it is certain that some part of its present population could not be kept going if we lost our position in international trade. No sensible politician would be in favour of a line of action that might involve this — even if he thought good might come out of it on balance in the future. That is the point: it is choosing a certain evil for a good that is, after all, hypothetical. And to accept odds of that sort is what we condemn as fanaticism.

Fortunately, another answer to the question of capitalist stabilisation is possible within the terms of Marxism: that of Western Socialism. It sees that capitalism is changing in character, and as it changes, is becoming capable of socialisation in many ways. There is at least as much evidence in favour of its view of the situation as of the Communist, besides giving more ground for hope. If there are conflicts of interest between one capitalist country and another, there are also bonds of common interest. They are more bound together financially and industrially than before; they are more indispensable to each other. World finance is on the whole an influence for peace; and even the great basic industries tend to get more internationalised, and to deal with their problems in common, as in the case of the European coal consultations at Geneva this year. On the side of the working class, admittedly the weakest, there are Trade Union contacts with other countries and international federations. The same applies to the relations within imperialism, that is, the case of a capitalist country and its dependency. There is the way of concession, governed here too by the degree of common interest that obtains. Trade is not all exploitation; it is, and is likely to remain, as much to the interest of the colonies to trade with us as it is for us to trade with them.[1] Lastly, it is within the capitalist State that there is the direct advance of socialisation. All kinds of facts come to mind: that one-tenth of the total property in England is public property; the tendency of undertakings of a national character, like the Bank of England and the five great banks, the railways and transport services, to develop the semi-socialist character of the " public concern "; the social legislation of the modern State by which wealth is to some extent redistributed for the end of the whole community. Now all this changes the conception of the problem; and a good Marxist ought to admit that different circumstances require a different solution.

Western Socialism then would answer that it is possible to carry through the transition from capitalism to Socialism by many and various methods; and that the end of Socialism is the same in both cases, the classless State and the community in control of production and distribution.

And so the fact that world revolution has subsided, which has given such difficulty to Bolshevik theorists, is easy enough of interpretation in the light of these facts. Capitalism has to some extent succeeded in stabilising itself — at the price of admitting Socialist elements into its structure which will ultimately supersede it. The Revolution has entered upon a new phase concurrently with these circumstances. During the early years 1917–22 it was primarily international: Russia, by a conjunction of circumstances, happened to be the terrain on which the Communist experiment was tried. But the figures of the early Revolution were not more Russian than European: Lenin, Trotsky, Zinoviev, Radek, Rosa Luxemburg. The Revolution is still international in its aims, but with a great difference; the revolutionary activities of the Third International are subordinate to the aim of building

---

[1] v. G. D. H. Cole, *The Next Ten Years*, p. 307: " Great Britain is fortunate in that her economic interests lie above all in raising the standard of life throughou the world ".

up Communist Russia. The differences between Stalin and Trotsky range over the whole area of Communist theory and practice; but this is the essential divergence.

There has been a most important change from the monistic internationalism of the early days. It was really too much to expect that the Socialist movements of the world could be run from Moscow, any more than from London, by a sort of Vaticanism of the working classes of the world. There has been a movement of withdrawal in all European countries. In France and England the party has lost ground by constant dissensions, to such a degree that with us it has lost whatever political importance it had, and no longer serves even to embarrass the Labour Party seriously. Anyhow English Communism, like English Catholicism, is bound to be more ultramontane than anywhere else, just because it is such a hopeless fraction in a vast profane environment.

An important point emerges that this new development may make it possible for the Socialist movements of the world to restore their broken internationalism, on the basis of Jaurès's ideal: " *la libre fédération de nations autonomes* ".[1]

There remains the outer fringe of imperialism. One can observe the disturbance effected in the world by the October Revolution spreading outwards to the margin, as in the case of the French Revolution which radiated its influence from Paris into the Netherlands, Germany, Italy, Spain in turn, and ultimately reached Greece and South America in the eighteen-twenties. It may be argued that Moscow is developing the one-sided internationalism of the Eastern peoples in turmoil, the Chinese and the Indians, having withdrawn from European hopes somewhat. The question is, have they any more chance in the Far East? It depends on two factors mainly: on the internal development of the Revolution in those countries, and on the policy of the imperialist Powers. With regard to both, there has been not much encouragement for Communists in the events of the last two years. In China they have had a severe setback; the popular movement has

---

[1] For a discussion of Jaurès's ideas on this point, *v.* the brilliant little work of M. Gaëtan Pirou, *Les Doctrines économiques en France depuis 1870*, p. 62.

got into the hands of the middle class and become identified with middle-class nationalism. Similarly in India, the real danger to the British Raj does not come from the proletariat and Communism, but from the middle-class demand for economic and political independence. The latent appeal of Communism is being countered by a policy of concession to the latter: this may be also a danger to the working classes of the East; but it may not be any the less successful in resisting Communism. The policy of agreed concession is likely to advance under the influence of Labour and Socialist governments in the West. Moreover, one cannot decide by theory alone that it is impossible for amicable and equal relations to develop within an imperialist system. Communists take it too much for granted that they cannot; whereas the best imperialists (of the *Round Table* in England, for example) are actively engaged in working out ways by which they may be achieved.

Certainly Lenin, as a realist, understood the prime difficulty of extending the Revolution to the Western countries.

In his articles upon the nationalist question and in his speeches at the congresses of the Communist International, Lenin more than once maintained that the victory of the world revolution was impossible without the union, without the revolutionary coalition, of the proletariat of the advanced countries with the oppressed people of the enslaved colonies: [1]

— surely an unlikely event? The continuance of the colonial relation by which the English working class is enabled to live on a higher economic level is at least as probable.

These, and similar difficulties for Communist theory and practice, are due to the excessive rigidity of their theory: one finds constantly that the analysis is on the mark, and is then pressed to conclusions which are falsified by the obstinacy of the world of facts. But as regards nationalism — the simple dualism of class war or national war? — one doubts if even the analysis is subtle enough. In its insistence upon class conflict it fails to recognise the actual four-squareness of the

[1] Stalin, *op. cit.* p. 186.

conflict: that there may be a common interest between capitalists and workers in one group as against another, even if at the same time there is a conflict of interest within the group. This over-simplification has led them to neglect, and in practice to be defeated by, the fact of nationalism. The Communist view is that it is largely a middle-class manifestation, and the most potent deflector of proletarian discontents that capitalism has to rely on — now that religion is no longer what it was in performing that function. But there still remains an element of good in the national cultures of the world, and one wonders if the culture of a unified internationalism would be really desirable. The fact that nations are so strongly rooted implies that they answer a profound need in human society: they may in the end be transcended as a form of social organisation, but meanwhile it is a mistake to deny their strength, which is a fact and inescapable. In so far as Stalin's views on nationalism are nearer realities than Trotsky's, Communism in the future is likely to be more accommodating: which is all to the good, for we would all prefer to attain freedom our own way.

The controversy among Communists over the question of nationalism may mean the loosening of the bonds of their theory; but, as things are, it is an illustration of the difficulties due to a superfluous rigidity of system, and to the exaggerated importance they attach to the notion of conflict. They are convinced — as regards the choice between nation or class, for instance — that conflict of one or the other kind is unavoidable. Marxism is right in its insistence upon the conflict of classes as the most powerful motive-force in society. But it gives these conflicts too homogeneous a character; and there is the further point, that they are not by any means always expressed in force, and that when their character is understood, it is possible to guide them to solutions based on compromise where the appeal to force is avoided. Indeed, political activity is of this nature; and in one sense, revolution only appears when this has broken down. Sometimes in history the revolutionary change is made without even breaking the continuity of constitutional forms, as in the case of the Reformation in England. And there is a strong case, on an historical argument, for holding that in England in the future what was achieved by the Russian Revolution will be attained by other and less primitive means. For the method of revolution is primitive; it is precisely as civilisation develops, and as society becomes less immediately dependent upon its economic basis, that the struggles of society can be directed into the channels of political action. It is at least possible that in the end a struggle may be avoided between what the new Russia stands for and this country, because the supremacy of the working class which is the common Socialist aim, will have been attained, and in our case without resorting to the dictatorship of the proletariat.

This is a much more desirable, a more economic way. On the question of force, the ground has been gone over in Trotsky's earlier book, *Terrorism and Communism*. He has a case; but it looks less well now, sixteen years afterwards, when he has to admit that the force that is justified in the interests of the whole lets loose the force that is not. This is the fate that dogs all revolutionaries; and Trotsky has come across it, in what he complains of as the political cynicism of Stalin. Even without questioning the rightness of the dictatorship of the proletariat for Russia, one may fairly wonder if Trotsky finds the method of dictatorship so superior to democratic methods, in the circumstances of opposition? Surely it is a more advanced state of politics, where Government and Opposition can coexist without endangering the conduct of affairs or the constitutional structure? The fact of an Opposition growing up inside the Revolution is itself evidence of the inadequacy of its institutional development.

Or one may turn the tables on the theory of the dictatorship of the proletariat and apply to it the solvent influence of historical materialism. Then it becomes no political formula of universal application, but the appropriate product of conditions in Russia on which Lenin was reflecting.

Or again, there is the case of the dialectic of historical materialism to illustrate the undue rigidity of their theory. The application of the dialectic to history marked a great

The Theory and Practice of Communism

advance on previous conceptions of historical processes; it gave a convenient form in which to include the coexistence of contrary tendencies, and out of their struggle the emergence of a synthesis carrying the movement forward. As a form it may be the best under which to conceive the general movement of history. But it is a *form*, and does not in fact express all the convolutions and recessions of society in action, the intricacy of historical events. And it is especially open to criticism when it is used as the basis for the prediction of revolution as the inevitable synthesis. Marx did not always hold it as inevitable, or a necessary process — in England, for instance; and even if he sometimes did, it is no reason for regarding his authority like Jehovah's: one needs to regard it critically, if we are to dethrone the gods and advance the age of reason.

Yet, in spite of its weaknesses and exaggerations, one comes to the conclusion that the general outlook of Marxism is a better and more appropriate approach to understanding modern society than any other. The ways and methods of Communist politics, especially in this country, have not my support; but I regard the body of Marxist theory as on the right and necessary lines. It was the first to grasp that the problems of modern industrial society have to be regarded economically if they are to be understood, and that this is the clue to other forms of social activity as well. In short, it is Marxism that is the intellectual system relevant to our time; and those who have no understanding of it do not understand the roots of modern society.

In a previous essay [1] I suggested reasons for the curious neglect of Marxism in the country of its adoption and formulation. Certainly it is extraordinary that the English Labour Party should have been so unaffected by Marxist thought. It is only the less surprising in that Fabianism gave it something of an economic conception of society, but practical and administrative in character, not analytical and revolutionary.

It is impossible here to go into all the characteristics that make Marxism as a body of thought so appropriate to our time;

[1] Pp. 224–6.

one may take two only of its most general characters: its combination of the assumptions of an irrational world with a rationalist outlook; and secondly, its essential internationalism.

Major Walter Elliot, in a vivacious little book on *Toryism and the Twentieth Century*, argues that the strength of Toryism is that it is based on the irrational in man. It is an interesting piece of evidence. And he is content to leave it at that. But there is no point in understanding how irrational men are, if the process of understanding is not a reliable, *i.e.* a rational one, and if the result of understanding is not to lead to rational ends. The rational must guide the irrational. And nowhere is the combination better made than in Marxism, which sees how irrationally men move in society, instinctively reacting to their conditions, impersonally in the mass; and then finds the reason in the action, makes it more attainable by making it explicit and avoids clashes with the inevitable in the process. Your freedom is to know best how you are bound.

The international outlook of Communism is essential to it, and though this is where it most comes into conflict with the rest of the world, it is at the same time its strongest point. It must be admitted that the internationalism of Western Socialism has made a poor and ineffective showing since it was broken and torn asunder by the war. On the threshold of that tragedy, it lost the leader who most typified its ideals: Jaurès. But the war split it everywhere, in some countries permanently, here temporarily. The secession of the Communists and the formation of the Third International stultified European Socialism; the Russian Revolution became as much of a divider in the Western countries as the French Revolution was with the Whigs in England. The War and the Revolution are responsible for the postponement of a Socialist Europe probably by a generation.[1] In the post-war world, Communism has developed a more ultramontane internationalism and has become identified with intransigent proletarian movements. This has had the effect of making a deeper rift in Socialism; and events in Russia have made the Communists themselves

[1] Cf. Halévy, *The World Crisis*: "Towards Revolution".

as fissiparous as the early Christians. (Still, that did not prevent the one true Church from emerging in the end.) Communism, even on Stalin's interpretation, holds essentially by its international aims: the programme of Socialism cannot be completed in a capitalist world. It may be that the Third International, like the Catholic Church after the Reformation, has achieved an all the more effective unity on the more limited territory left after the rift. But it is not so limited either, and deserves the credit for being the one political internationalism which has eliminated the barrier of race.

The fate of Communism depends on the crisis in Russia; and yet, whatever eventuates from it, the achievement is already of enormous importance and must leave permanent marks. There is not much doubt that on the whole the standard of living, in spite of the losses in the turmoil, is slightly higher than before the War. What is more significant is that a new society is being built up, a new attitude of mind for a whole people: where formerly their mentality was characterised by passivity, mysticism, and fear, it is now under Communist training marked by optimism, activity, and a stark positivism. The proletariat is in power: to maintain it is the end of government. An ironic sense of humour in politics has dictated that only the bourgeois and ministers of religion should be disfranchised. All observers of conditions in Russia are agreed as to the irresistible force generated in this great people by the Revolution and by the direction of their social life under Communist ideals.

We are concerned here more with the probable effects upon politics and culture. It is evident that as yet we are only at the beginning of the Marxist influence upon society and its life. In Russia the forms of capitalist society no longer obtain, any more than its pretences and its prejudices: conventional morality dependent upon a semi-religious view of society is no longer officially supported, any more than religion is; in fact the latter is strongly discouraged by the propaganda of the Government. And at the same time as there is moral freedom, there is an active vein of positivist puritanism among the politically-minded, which is certain to be the driving force among the new generation.[1] Intellectually, all kinds of influence are brought to bear against " defeatism ", mysticism, and in favour of a positivist outlook and class consciousness. But we have seen nothing in England of its influence upon art and literature, very little of its film technique and the drama, and but little more of its effect on science and thought. Yet the extension of Marxist views, outside of Russia as well as inside, is bound to have the most profound influence upon all intellectual currents, upon literature as well as science, upon criticism as upon economics, upon history and philosophy. Who knows what it may yet create?

For Marxism is something more than a body of theory, and Communism than a political system. It is a Renaissance of its kind : it has the vitality of a new humanism. Its output in theory and its results in practice are astonishing. Yet in the history of civilisation it is the expression of values by which a living movement is justified. It has to give new impulses to art and the mind's life, as well as to science and the practical world. Perhaps it has already added to the values of men's lives in Russia. There they may be on the verge of such a movement, a new form of society; or the experiment may fail, and the waste prospects of capitalist civilisation close in upon us again.

[1] This has been completely borne out in the years since, to such a degree in fact as considerably to impair the moral freedom of the earlier years of the Revolution. At the same time the persecution of the Church has come to an end, and there has been a *rapprochement* between the Russian State and the Russian Church. [1945.]

XX

MARX AND RUSSIAN COMMUNISM

(1) [1939]

AT last, thank goodness, we have reached the time when Marx and Marxism can be written about in a reasonably intelligent way. No longer the dreary, squalid little manuals written by insufficiently instructed persons on the one hand — the uneducated writing for the uneducable — nor, on the other, the blank un-understanding of the academically superior. At any rate, if such works continue to appear, we do not need to read them. With Cole's writings and those of Sidney Hook, Lindsay, and Carr, there has at last come into existence a small literature on Marxism in English — quite enough to go on with — which one can read without contempt for the ignorance of some or irritation at the obtuseness and self-sufficiency of others. That is to say that this recent work is sympathetic yet critical, understanding, and intelligent; it exemplifies the same intellectual standards as would be applied, say, to John Stuart Mill or Darwin or any other great figure in the history of thought. Marxism is neither swallowed whole uncritically as so much revealed and unquestionable truth, nor condemned outright as so much nonsense. Why should it be either? Either attitude is extremely silly.

Mr. Berlin's attitude to his subject is exemplary, and on the whole it is the best introduction to it that we have.[1] His little book has several advantages: it is not merely a biography, but gives as much space to Marx's thought, and, for once, from the point of view of a professional philosopher. The book is not perfect: who could write a perfect book on Marx of all people? The proportions are a little out: much more space is given to the early than the later Marx. That is right, on

[1] *Karl Marx*, by I. Berlin.

the whole; but the later economic writings, with *Capital* at their centre, are inadequately dealt with: there is a lacuna here. Mr. Berlin is more interested in the formation of Marx's doctrine, and a fascinating job he makes of it. The great quality of the book is its absence of *parti pris* and its completely impartial and objective approach. In consequence it makes Marx intelligible, both as a person and as a thinker; and without any undue admiration — in fact, Mr. Berlin is a little too cold about him, admires the G.O.M., extraordinary and unattractive person that he was, rather insufficiently in my opinion. One can admire even where one does not like.

This is no place to argue the familiar, the stock pros and cons of Marxism: I wish to draw attention to what I think most original in what Mr. Berlin has to contribute. He realises, as so few have done, the fundamental rationalism of Marx's outlook. Not in the sense of liberal rationalism: Marx did not fall, no one less, for the most superficial of rationalist fallacies, the view that people in the mass are essentially reasonable, will respond to rational persuasion, etc. We all know how untrue that is. He held, all the same, that "a man's actions were inevitably guided by his real interests, by the requirements of his material situation". But what if, after all, they are not? or are only partially or discontinuously so? Marxism has certainly opened our eyes to the way in which men's behaviour is motivated by interests, but it does not allow for the extent to which it is the result of sheer foolery or stupidity. What is so astonishing in the contemporary world is not so much that some men (and groups) are motived by their interests, as the way that others are not: not for the old liberal-idealist reason that they are above them, but the only too realist one that they are so stupid that they do not know what their interests are. When you universalise that degree of idiocy over the map of Europe and beyond, you arrive at the contemporary scene. It almost seems that they would prefer to take the path of destruction instead of that of rational calculation and both self- and group-preservation.

Marx and Russian Communism

To take a plain yet very important example for Marx's politics: he placed all his hopes in the working classes, yet nobody understood better what fools they were, " infinitely gullible, obstinately loyal to the agents of their own worst enemy, who deceived and flattered them only too easily to their destruction ". Look at the way in which the masses have followed the Hitlers and Mussolinis, the Baldwins and Chamberlains in the past few years! There is a discrepancy between Marx's pessimistic view of human nature, which was, on the whole, well justified, and his political optimism, which looks as if it were premature by a century or two. The reason intellectually, no doubt, was part of his legacy from the awful Hegel: the assumption that the historic process is itself rational and progressive. But suppose if it is not? It is really much too simple to assume that " history does not move backwards or in cyclical movements: all its conquests are final and irrevocable "; that the bourgeoisie must necessarily be defeated by the working class, as the feudal nobility was by the bourgeoisie; that the victory of the working class will mean the abolition of all classes, etc. etc. It is really all very naïve, making history march *comme sur le papier, qui souffre tout.* Whereas history is, in fact, a much more complicated process. Over long periods there is retrogression; civilisations have disappeared; societies have been defeated as often as they have won; men have failed, leaving the ground encumbered with the ruins of their hopes, or vanished off the map without leaving a trace to show that they could not solve their problems. If it would be an advantage to Professor Toynbee to have read a little more Marx, it would have been no less an advantage to Marx to read Professor Toynbee.

What is more important for contemporary politics, it is this that explains why Marxism has been so caught out and its hopes postponed by the reaction of the decades in which we live. We do not need to insist here that its analysis of society is more penetrating and apposite than any other before us; but it is insufficiently sceptical, too rigid and formalised, too Hegelian still. It must learn something of the wisdom of Hume. Even so, the thought crosses my mind that in these degenerate days people are perhaps unworthy of Marx and Marxism; they understand neither him nor it nor anything very much. It is much too intelligent a system, even as it is, too enlightened, too reasonable, too right. Now perhaps in the twenty-first or twenty-second century people may see the point of it and it may come into its own. But a great deal of retrogression may also take place before then.

(II) [1936]

Of all the books on Marxism, this is the most important that has appeared for some years.[1] And that in itself gives it considerable importance; since there is nothing more remarkable than the way in which the issues central to Marx's thought have become the chief intellectual issues of our time. Professor Hook says, rightly, that

the phases through which Marx's thought developed recapitulate the difficulties faced by critical minds today when confronted with the Marxian position. There is hardly a doctrine urged against it, from the latest variety of ethical idealism to the newest twist in psychological self-interest theories, which does not have its precursory expression during the years when Marx fought his way to the philosophy of dialectical materialism.

Professor Hook is already well known for what is perhaps the best general introduction to Marxism; in this volume he traces in detail, and with careful documentation, the various steps by which Marx reached his position, at first by way of reaction from Hegel, through controversy with the young Hegelians, with Arnold Rüge, Max Stirner, and Moses Hess, through criticism of Strauss, Bruno Bauer, and finally of Feuerbach.

At last Marx is receiving the treatment proper to a classic, as it might be Kant or Hegel himself. Professor Hook's method is critical, expository, scholarly; at every point he cites Marx's position in his own words, along with that he is criticising. The result gives one an extraordinary impression of the comprehensiveness of Marx's thought, its consistency, brilliance,

[1] *From Hegel to Marx*, by Sidney Hook.

and good sense. All the world knows of Marx's contribution to politics, economics, history; but very few, even of the intellectually cultivated, realise how his position developed as the result of the closest criticism of philosophical idealism. When one considers the range and power of his mind, it is hardly extravagant to think of him as the Aristotle of modern thought, as indeed the whole cast of his mind, and his intellectual sympathies, were largely Aristotelian.

This may be observed in every field, whether in his criticism of religion or of idealism, in politics or history — his insistence upon society as the proper framework of reference, to which every phase of human experience or knowledge is to be related. This constant reference of everything to the whole of which it is part, is what Marx has in common with Hegel; though where, with the latter, the whole is the self-evolution of the Absolute, with Marx it is concrete human history, the evolution of society and its interrelations with the physical environment. Marx is the most historical of thinkers; he may almost be said to have been the thinker who discovered history. It is extraordinary how Hegel, magnificent and monstrous thinker that he was, though himself the first to appreciate the evolutionary process, had so static a conception of it. With Marx the historical process is so much in the foreground that one wonders whether all this relativism does not lead in the end to historical scepticism.

There is a genuine difficulty here; for over and above the social conditions forming a certain body of thought — and nobody is more illuminating than Marx on the question of origins — there remains the question of validity. It often looks as if he is not interested to deal with the ideas in and for themselves, so bent is he upon analysing the conditions giving rise to the theory. This is apt to lead, as indeed has been noticeable in the subsequent Marxist tradition, to scepticism with regard to truth, or else a too facile pragmatism, dangerous in action because of the absence of rational tests. Take Marx on ethics (the words are Engels', but the position is Marx's):

We maintain that up to the present time all moral theory has been the product, in the last analysis, of the economic conditions prevailing within given societies. And since society has hitherto developed through class antagonisms, morality has always been a class morality. . . . We have not yet advanced beyond class morals. A really human morality which transcends class antagonisms and their memories will not be possible until a stage in human history has been reached in which class antagonisms have not only been overcome but forgotten in practical life.

It would be easy to challenge the optimism, the underlying rationalist assumptions of this: in more senses than one, Marx and Engels were on the side of the angels. But the main point is that though much of the content (and still more of the form) of ethics changes with changing conditions, may it not be that something of its content relates to fairly permanent and stable elements in human nature, through all the changes of human history? Marx tells us, quite rightly, that there is a history of standards themselves; but is there not a standard of standards? Is there not a certain continuum of human experience to which the variations may be related, and that saves us from that utter scepticism to which historical relativism would drive us?

The point may be made clear from Marx's criticism of religion. Strauss regarded religion as the product of the myth-making consciousness of man; Feuerbach related religion at every point to man's nature, interpreting religion, in a curiously modern, almost Freudian way, as the projection of man's wishes, " the dream of the human mind ". Marx criticised both for not realising how dependent religion is upon the particular social conditions which give it its character at different times. There is much that is brilliantly suggestive in his historical approach. Anyone nowadays ought to be able to appreciate the extent to which medieval Catholicism with its carefully graded hierarchy and formal rites was the product of a feudal age, while Protestantism developed along with commerce and bourgeois standards, simpler, time-saving, cutting down waste, self-reliant, individualist. And Marx certainly had hold of the right end of the stick, politically speaking, when he wrote of the identification of religion and political reaction in nineteenth-century France — " the mortgage held by the peasant on the heavenly estates, guarantees the mortgage held by the bourgeoisie on the peasant estates ". After all, Tocqueville thought much the same thing, and saw, from his point of view, that it was good.

But Marx goes further than this to deny a specific history of religion as such; he would allow only the history of general cultures of which religions were but fragmentary aspects, related to their particular social environment. He regarded religious beliefs as arising from social maladjustments, and tackling the social maladjustments as the way to curing the beliefs. No doubt he was right as against Strauss and Bruno Bauer in insisting that specific criticism of religion could only be satisfactory when completed by a comprehensive criticism of society. But it is a mistake to push historical relativism too far; does it not appear that there is an element that is common, and historically continuous, throughout all religion, whether it be worship, or propitiation of the unknown? It may be that it appeals to an element in human nature that persists through changing historical conditions.

These points illustrate what is a general criticism that may be made of Marx's position, that it does not allow sufficient autonomy to other aspects of human experience, with their appropriate disciplines, than the social. His own social approach to everything is abundantly justified by the illumination it has thrown into every corner of knowledge; and in his own time, and right up to our own day, it has served to correct the emphasis which in most men's minds leans in the other direction. But not all the problems of existence are material, or even social; spiritual questions are no less real, after their kind, and in their appropriate realm. The problem is how to relate them both; a more pluralist view than Marx had would seem to be necessary to comprehend the whole of life.

Occasionally Marx proves himself more liberal and less one-sided in this respect than others; for instance, in regard to literature (like Lenin later who was much more sensible in the matter than the doctrinaire young Communists), Marx was much more catholic than the young Hegelians. The latter, with Rüge at their head, produced a regular polemic against the Romantic School. It was of course based on social and political grounds. But it often happens that what may be disapproved of politically leads to an admirable achievement in art; *Murder in the Cathedral* is a case in point. Aesthetically, the romantic reaction led to new and desirable developments, to further possibilities and achievement in the arts. It is necessary to distinguish between social and aesthetic criticism, as Marx did in coming to the rescue of Goethe from the young Hegelians, discriminating between the political reactionary and the poet who was a genius.

It is unnecessary here to point out how much less doctrinaire Marx was in the realm of politics than many of the young Hegelians, particularly the " True Socialists ". The " True Socialists " were almost uniformly true idiots, refusing any cooperation with the general democratic movement of the time; rather like the I.L.P., lost in the contemplation of their own virginity, when what was needed was the assertion of a little political sense. This Marx had in abundance, and an acumen and originality amounting to genius. An analysis of the German bourgeoisie and of its utter political bankruptcy, made in the eighteen-forties, holds good a hundred years later, such is the penetration of its judgment; it is with a start that one realises one might be reading of the Weimar Republic after 1918:

> Through the July Revolution ... the perfected political forms of the bourgeoisie were imposed on Germany from the outside. Since the German economic relations had not even begun to correspond to these political forms, the bourgeoisie accepted these forms only as abstract ideas, as principles valid only in and for themselves as pious wishes and phrases.... They consequently handled these principles in a much more moral and disinterested manner than other nations. That is to say, they made their highly peculiar stupidity absolutely valid and remained unsuccessful in all their efforts.

But this is not the place to deal with Marx's political and economic thought. Professor Hook promises us a further series of studies on its development in relation to the French and English writers who so greatly influenced it. In this volume he has opened up a fascinating field of investigation, all the more exciting because few of Marx's early works have been at all adequately translated. Professor Hook would put us under a further obligation if he would later translate some of these works for us, particularly the *German Ideology* and the *Holy Family*. He would certainly deserve, if he did not receive, the thanks of the English-speaking world.

(III) [1937]

Professor Carr has given us, what is rare, an historical biography that is also a contribution to history.[1] He has had a first-class opportunity in the life of Bakunin, and he has made the most of it. For there have hardly been any biographies of Bakunin, of any value, in any European language, let alone English; and in consequence his life and career have remained a lacuna in the knowledge of most of us even who are interested in the history of Socialism. Again, Bakunin's peripatetic revolutionary activities in Russia, Poland, Germany, France, Switzerland, in Berlin, Brussels, Paris, Dresden, London, Stockholm, Geneva, Bâle, Lyons, his imprisonments in Austria, in the Peter-and-Paul fortress at Petersburg, his escape from Siberia via Japan and U.S.A. which brought him world fame — all this brings his career into touch with the main stream of nineteenth-century history at innumerable and often unexpected points. Or, to vary the image, Bakunin's life is so much of the stuff of the squalid and tattered underlinings to the respectable surface that the nineteenth century presents to us in the text-books.

Yet the history of Socialism and of revolution, which by no means occupied the centre of the picture in that age (except for the one solitary year 1848), has become of increasing importance in our time now that we can see to what those subversive activities, those romantic gestures and fantastic careers, the ridiculous capers and incredible hopes were leading. It is not so much the respectable Queen Victorias and Gladstones and Peels, or the Newmans and Mannings and Pio Nonos, or

[1] *Michael Bakunin*, by E. H. Carr.

The End of an Epoch

the Nicholases and Alexanders, or even the Bismarcks and Moltkes of the nineteenth century who have a message for this; it is the outcasts and exiles who matter more now, the unrespectable, Marx and Engels, Proudhon and Bakunin and Lenin. Professor Carr is fortunate in his choice of subject.

And he has written with feeling for the background in place no less than in time, which makes, in spite of the book's length, for easy reading. The picture that is drawn of Bakunin's early years and family life at Premukhino, charming as it is, reminds one of a Turgeniev interior or landscape — very literally, for Turgeniev was a friend of Bakunin's and it is well known that the character of Rudin was modelled after him. Bakunin's career took him to many odd places off the beaten track, which Professor Carr's knowledge of the nineteenth century never fails to illuminate; in addition to Russian history, there is Poland and the different forces in the Polish nationalist movement, the situation in Sweden and Finland in the 1860's no less than Paris in 1848. Professor Carr has followed with admirable patience the vagaries of this exasperating career, the generosity, the devotion, the indiscipline, the endless squabbles, the absurdly sanguine programmes, the atmosphere of unreality that pervades it all, the debts, the broken promises, the squalor. Not a word of protest, hardly of criticism; he leaves the story to make its own impression, the reader to draw his own conclusions. This has its disadvantages no less than its advantages.

For one thing, if one may venture a word of criticism of so excellent a book, it sometimes leaves the reader in the dark as to *what* Bakunin was actually doing. One has the impression of a great deal of sound and fury, but *what* was he doing? one asks. One knows very well what Marx was doing contemporaneously — editing or contributing to papers, studying hard, writing books, engaged in secretarial work for the revolutionary cause, intriguing. Perhaps our difficulty is due more to the subject of the biography than to its author. Bakunin was so very Russian; most of his time he seems to have spent in smoking and talking — I suppose it would be called propaganda.

Marx and Russian Communism

He certainly created a legend, and perhaps this is his one enduring achievement. Again this may serve to explain a similar failing in the book: it does not make clear *why* Bakunin thought what he thought, the intellectual processes by which he arrived at his position. Perhaps it does not matter so very much, for with Bakunin, more than with most intelligent people, his intellectual position was but a rationalisation of his impulses. But the book is weaker on the intellectual side, as opposed to the purely historical, which is its strength.

What is it, after all, that Bakunin's position came to? It may be described in the words of the uneducated tailor, Wilhelm Weitling, which had such a decisive effect upon Bakunin's mind: " The perfect society has no government, but only an administration, no laws, but only obligations, no punishments, but means of correction ". This may be all very well as an ideal, but since it rests upon a total misconception of human nature it is difficult to see that it has any importance whatever for politics. Bakunin seems to have attained to greater sense in the last year of his life when he wrote: " To my utter despair I have discovered and discover every day anew, that there is in the masses no revolutionary idea or hope or passion; and where these are not, you can work as much as you like but you will get no result ". It seems a pity that it should have taken a whole lifetime, and such a devious route, to arrive at that much sense. This being so, one sees how right Marx was to insist upon discipline, order, authority. And it seems providentially right — if one may use the term — that it should be not the great Russian, but the German, who should have so affected Russian history. For Marx provides just what the Russians stand most in need of; they have enough Bakunin in their composition, and too much, already.

(IV) [1941]

Mr. Wilson's new book has been received with the attention and respect that everything he writes deserves.[1] Perhaps this

---
[1] *To the Finland Station: A Study in the Writing and Acting of History*, by Edmund Wilson.

review may then more usefully concentrate on the critical rather than the appreciative aspects. Not that one is without appreciation of his effort or of the qualities that he brings to his work. He is one of the two or three most interesting critics of literature writing in English. And when he comes to the less agreeable subjects, the more stubborn material, of history and politics, and especially to the dreary tract of Socialist history and thought which he has chosen to investigate in this book, he brings to it the faculty of making even that live and interesting again. No mean achievement when you consider the weary library of books by the fourth- and fifth-rate devoted to the subject — most of which, for my sins, I seem to have read — on Saint-Simon, Fourier, Proudhon, Robert Owen, Marx and Engels, Lassalle, Bakunin, Lenin and Trotsky, and so on *ad nauseam*. Mr. Wilson brings the equipment of an educated and sensitive man to the consideration of the thought and work of these men. He makes them human, real, live men once again, as real and moving as, say, William Pitt or Nelson or Newman or Swift; instead of the unintelligible sawdust and cardboard of cheap Socialist publications for the populace, of which Socialist literature too much consists. Mr. Wilson has style and distinction of mind. He is a cultivated man: it is such a relief. When he chooses to turn his attention from pure literature to the difficult subject of the interactions of history and politics, the study of history by Socialist and near-Socialist thinkers as a basis for action, for bringing about changes in society, he is bound to have something interesting to say.

He begins with the historian Michelet, with his passionate and unbounded sympathy for the People; for him the People took the place that the Proletariat held for subsequent Marxist writers. (Michelet, astonishing as it seems, loved the people; the fact that he came from them was no excuse: it was an emotion of which Marx and Engels, at any rate, cannot be accused.) Michelet expressed all that side of the vital sympathies of the French Revolution, and did impress himself upon his century with his *History of France*. Then we are shown the progressive disillusionment with Renan, Taine,

Anatole France. Mr. Wilson is not quite fair to the last, who did put up a manful struggle for progressive causes and for justice in the Dreyfus affair. But really, why should they not have become disillusioned when you consider the deplorable record of their people in their time: the fatuous and unsuccessful Revolutions, the disgusting and brutal Reactions, the disillusionments and disasters of 1830, 1848, 1851, 1870, 1914, 1918 . . . etc.? Anatole France may well have broken down one day and confessed that at heart he was the unhappiest of men, when he reflected upon human desolation: the marvel is that he should have contrived such a gay, courageous mask to wear on the outside. All these men, even the most sceptical of them, Renan, were on the side of better things and did what they could to advance them — to find themselves sabotaged in the end by the incurable folly, the stupidity and barbarism of men in the mass. One can hardly deplore their disillusionment: surely these men were right as against the monstrous, the fanatical idealism of Marx and Engels, Lenin and Trotsky, the optimism ceaselessly betrayed and shown to be mistaken, and as childishly renewed? No wonder, as Mr. Wilson tells us, in the Marx-Engels correspondence " the word *Esel* seems almost to become synonymous with *human being* ". They were only too justified. (In politics the word human = fool; the animals after all are by definition incapable of being fools.)

Beside this petering-out of the bourgeois tradition of hope coming down from the French Revolution, Mr. Wilson sets the rise of Socialist thought from its utopian origins to maturity with Marx and Engels. He shows how Marxism did give a closer understanding of the forces at work in society than before, and how Socialists took it into the field of practical political action. With not much success, it should be added, until Lenin came along, and at his entry into Russia at the Finland Station in 1917 " stood on the eve of the moment when for the first time in the human exploit the key of a philosophy of history was to fit an historical lock ". No doubt his Marxism helped him to understand his opportunity; but I should say it was his opportunity that was infinitely more important, and, Marxism or no Marxism, the fact that he had the political *nous* to exploit it.

Though Lenin is Mr. Wilson's hero, the Finland Station is not really the moral, but only the terminus, of his book. His own disillusion with Marxism, his realisation of its inadequacy, on which he speaks for the most intelligent of his generation who have passed through the same school, would logically point to another moral — that brought out in the burlesque on Stalin in the Appendix. He says in an aside that " we are living at the present time in a period of the decadence of Marxism ", and probably quite rightly from the intellectual point of view. The paradox is that most ordinary people do not as yet realise how much was to be said for it, the strength of its case or its achievement. To us, however, its inadequacies are all too apparent. He does not specify them all, nor are they equally important. But he is clear that the Dialectic is a religious myth, and we all know how prone the Germans are to myths — their chief contribution, it would seem, to humanity. The fact that the Dialectic comes from Hegel should be enough, I should have thought, to warn all sensible people from swallowing it whole. But, of course, *sensible* people never have swallowed it. It was merely the fanatics who insisted that it was essential to Marxism; and see where it has led them — into the worst dangers and disasters of mere pragmatism and political opportunism. The attachment to the Dialectic has played its own nefarious part in blinding the eyes of Communists to the criminal follies of Comintern policy and of Stalinism. The whole world is now strewn with the wreckage. This is what happens when you throw away rational standards and tests for action, and can no longer see the difference between principle and expediency, Truth and Dialectics, sound strategy and mere tactics. I have long thought that on the intellectual side, the Hegelian Dialectic should be discarded in favour of a more sensible, more cautious, less ambitious, and more truthful empiricism, the radical empiricism and scepticism of a Hume.

With Mr. Wilson's criticism of *Das Kapital* goes a good deal of Marx's economics: the confusion over the Labour Theory

Marx and Russian Communism

of Value puts the system *qua* system in question, though a good deal of penetrating analysis remains and much valuable history. But the role of history in these men's minds can be more profoundly criticised: as hypostatisings of their own ego. When Lenin says, " History will not forgive delay by revolutionists who could at once be victorious ", he only means I, I in the Future = Success. When Trotsky tells Martov and the Mensheviks, " Your role is played out. Go where you belong from now on — into the rubbish bin of history ", he only means his own rubbish bin. Trotsky found himself in the end relegated to the rubbish bin; but that does not mean to say that his protests were not absolutely right against the idiot policy of letting Communists and Social Democrats do each other in under the Weimar Republic, etc. etc. And that implies what is of greater importance still, that there are values and principles and standards of conduct outside the Dialectic. What is the point of sacrificing men's very lives for the sake of action in itself? It all depends on where the action is leading whether the sacrifice is justified. There is a kingdom of ends that is entirely ignored by the political pragmatism to which Marxism in our time has been debased.

Moscow is now having to pay for it. The tragedy is that so much is bound up with that bark, and if it founders, the hope of any European civilisation in our time founders.

It is clear that Mr. Wilson thinks all this, though he does not bring it out fully in his book. His book has an inconclusiveness, a tentativeness of touch and disparateness of method which weaken the treatment of this theme — the greatest of contemporary themes. Instead of being a study, it is, like all his books, a series of studies. He does not seem capable of the unifying, intellectual effort to write an integrated book — though no doubt such a book would require a tremendous effort and immense knowledge and research. However, it is abundantly clear that he is on right lines, that he understands what has gone wrong and where and why, and what is at stake. I find myself in entire agreement with him, and I fancy he speaks for a great many of us who have been through these years with our eyes open and not lost the capacity to think. If the light goes out here in Europe and the disaster comes that we have so richly deserved, it is consoling to think that there are a few choice spirits like Mr. Wilson in America to carry on the tradition of intelligent reflection on these despairing issues.

(v) [1931]

This is a very important book, perhaps the most significant study in the field of political institutions since the War.[1] And this is no mean praise at a time when the quality and amount of the work in the narrower field of political science, what with such books as Mr. Brogan's *American Political System* and Dr. Finer's appearing, has been so high, while political theory proper has languished in the coma that overtook it with the premature death of T. H. Green. Moreover, there is an element of the sensational in the performance, of which the Webbs are themselves conscious and not a little proud. It is a remarkable thing that we should have had to wait for the fullest account of the new social order in Russia from " two aged mortals both nearing their ninth decade ", as they describe themselves. They put down their presumption to the recklessness of old age.

In our retirement, with daily bread secured, we had nothing to lose by the venture — not even our reputation, which will naturally stand or fall upon our entire output of the past half-century, to the load of which one more book makes no appreciable difference. On the other hand we had a world to gain — a new subject to investigate; a fresh circle of stimulating acquaintants with whom to discuss entirely new topics, and above all a daily joint occupation, in intimate companionship, to interest, amuse, and even excite us in the last stage of life's journey. This world we have gained and enjoyed.

A good deal has been made of the disqualifications of the Webbs for treating this vast and new subject: their ignorance of Russia and Russian history, their very Englishness, the fixity and stiffness of their methods of social investigation. In reality, nothing is more remarkable than the flexibility of mind these two " aged mortals " have brought to their subject.

[1] *Soviet Communism: A New Civilisation?* by Sidney and Beatrice Webb.

Marx and Russian Communism

Little as they can have been prejudiced in favour of Russian Communism as against their own Fabian Socialism (had they not been made the targets of immeasurable abuse by Communists in the past twenty years?), they have ended up their prolonged study of the new Russia and its institutions very strongly in its favour. That is unmistakable. But it is not a disqualification; in fact it has helped their understanding of this new world to have been sympathetic to it; nor has it impaired their critical faculty, nor even, taking advantage of their being in a sense the grandparents of the Revolution, their readiness to read the Soviets a lecture upon a number of matters not to their liking, *e.g.* the "Disease of Orthodoxy", the dualism of Soviet foreign policy and the subversive tactics of the Comintern, the dogmatic character of Soviet atheism, their cult of hatred and abuse in politics. The qualifications of the Webbs for this great work of social investigation are obvious: a lifetime's experience in this kind of work, a profound and exhaustive interest in every aspect of human organisation in society, an unprecedented thoroughness, dispassionateness, and patience. After all, the variety of groupings open to men in society is not large, the associations they form are not so very different, though they may be inspired by different principles. It is true that the Webbs in their life-work have concerned themselves solely with English institutions; but this makes their comparisons of Russian forms of government with English all the more illuminating. Something is no doubt lost in the "feel" of Russian institutions, the sense of the Russian temperament and of the Russian past. One is inclined to comment on the careful explanations that are given of the reasons for the gap between what is planned and what is actually carried out, that the Russians after all are Russians, and have not ceased to be Russians for becoming Communists. And there is a simpler explanation for the ultramontanism of Communist orthodoxy, whether as regards theory or practice, dialectical materialism or the Third International, and that is the strain of ideological conceit in the Russian, the naïve and dangerous conceit of the clever child.

The rigidity of treatment, which is noticeable over tracts

of the book, is not so much due to this as to the Webbs' worship of social machinery. How they love machinery and institutions as such! If you want to find out all about the artel, and the incop, ispolkoms and kholkhose, oblasts, rayons, and selosoviets, read Vol. I. By the detail of their treatment, the Webbs have practically reduced this volume, except for the last chapters on the Communist Party, and on Dictatorship or Democracy?, to a work of reference. Intending readers of a weaker fibre than this reviewer would be well advised to read Vol. II first. That begins with something of an historical approach to the Revolution, which I should have thought a much easier road to understanding it and appreciating the magnitude of its task. Vol. I describes in all its fullness and "multiformity", the whole social, economic, and political system of Soviet Russia; or as the Webbs describe their successive chapters, the organisation of "Man as a Citizen", "Man as a Producer", and "Man as a Consumer", this triple collective organisation covering 170 million human beings, and one-sixth of the globe. It concludes with a treatment of the Communist Party as the force that guides and galvanises with energy this great mass. Vol. II goes on to discuss, much more interestingly, the issues of tremendous importance, social and theoretical, that are raised by these great experiments in society.

It should be said that there is hardly any question of importance to contemporary political science that is not faithfully dealt with, or at least touched upon by the Webbs; the practicability or no of a planned economy — they are perfectly *au fait* with the disputes going on amongst economists on that point and deliver many shrewd blows against the embattled professoriate of their own London School of Economics; the problem of replacing the motive of private profit by other sources of initiative under Socialism; the function of leadership in modern democratic conditions; the dangerous dichotomy between scientists knowing nothing of society, and politicians and humanists knowing nothing of the technical basis upon which our society is largely based. The very footnotes are ripe in the accumulated wisdom of fifty years' accurate and methodical reflection upon politics — and usually make sprightlier reading than the text. In short the book is a *Summa* of the Webbs' outlook on society, and bears much the same relation to the body of their work as Shaw's *Intelligent Woman's Guide to Socialism and Capitalism* to his.

Of the main issues raised by the book, there is room only to mention one or two. There is that of Freedom, of which so much has been made against the Webbs. On the dragooned uniformity which is exacted towards the Soviet system, they say rather naïvely, " It is only the calling in question of the fundamental principles of communism, or some aggressive criticism of theoretic Marxism — and of course any incitement to political ' faction ' — that is barred as ' counter-revolutionary ' ". This seems to under-state the position. I cannot but think that it would have been better throughout the book to make much more of the element of compulsion, which is necessary to keep the thing together. The Webbs are at great pains to point out how democratic the system is — they call it a " multiform democracy " or " democratic centralism " — and what scope it gives for the active participation of far larger numbers of people than the democracies of the West. And there is a great deal in what they say both on the economic and the political side. For example, the so-called freedom of choice which is exercised by the consumer under capitalism only applies to a minority of consumers; and they are quite right when they point out that it is this minority that capitalist economics is concerned with, assuming its freedom of choice to apply throughout the whole system. Similarly they are right in saying that the Soviet system politically provides as much constructive opportunity for the mass of the workers as Western democracy. They regard the power which works the system as " generated in the innumerable meetings of electors, producers, consumers and members of the Communist Party, which everywhere forms the base of the constitutional structure ". No doubt it forms the base, and there is a sense in which the people contribute to the operation of governing; but, surely, passively. Power and authority come from the top downwards. The whole institution of the Communist Party shows it to be so. The

The End of an Epoch

Communists govern and lead; the people, at most, collaborate, often inertly, or ineptly, or reluctantly. Why not say so? Stalin makes no bones about it. " In this sense, we may say that the dictatorship of the proletariat is, *substantially*, the dictatorship of the Party as the force which effectively guides the proletariat." The Webbs affect not to know what the dictatorship of the proletariat means; though the phrase bears its meaning upon the face of it. Or again, they are at pains to show that the system has the support of the whole country, or practically the whole. But what does that matter compared with the fact that the system expresses the *real interest* of the whole — even though there may be elements, as always, who do not know what their real interest is? The Communist Party is there to show them, to lead them, to guide them, to stimulate them, to attract them, to entice them; in the end, to *make* them. Again, though the Webbs appreciate the Communist Party as organising the vocation of leadership, they are wrong in estimating its role as unique. Every country has a governing class, only in Russia it is not rooted in economic inequality. It is none the less powerful for that.

On one issue, fraught with a terrible significance for the modern world, that of Nationalism, it will be agreed that Soviet Russia has been amazingly successful; and it may be that in the solution it has found lies its chief constructive contribution (so far) to political science. Communism has brought cultural autonomy along with a genuine social equality to all the racial minorities within the boundaries of the U.S.S.R. It is to this that the success of its federal system is due. And one consequence of its citizenship not being founded upon race, or nationality, or language, or colour is that there remains open to other peoples the possibility of associating themselves with the Federation of Soviet Republics. It is a potent means for the extension of Soviet influence, particularly in the East: witness what is happening in Outer Mongolia and Chinese Turkestan today. But in any case, can any rational being doubt how right this position is compared with the mania of nationalism that Fascism everywhere stands for?

Whether Soviet Communism is a new civilisation, as the

Marx and Russian Communism

Webbs argue, for all its achievements and promise, is another matter. They assemble a number of factors — the abolition of private profit-making, the establishment of social equality, the planning of production for community consumption, which mark it off from the West. But there is more to be said for regarding it, as Lenin did, as the westernisation of Russia — its industrialisation, the education of a backward people, the extension of a rationalist, scientific culture which gained its first victories in Europe in the seventeenth century. In other respects it is a more audacious and complete drive towards the fulfilment of the same trends that are at work in the West, towards social equality, the bridging and transcendence of class divisions, the emergence of the whole community into the foreground of political action. Beneath the many divergences and the bitter conflicts of our time, it may be that Russia and the West are moving towards the same ends and a similar society.

XXI

AN EPIC OF REVOLUTION: REFLECTIONS ON TROTSKY'S *HISTORY*[1]

[1933]

READING this book through from beginning to end — in itself no mean task, for there are three volumes of it and some thirteen hundred outsize pages — one feels that there is some consolation for the loss of such ability to the Revolution in Russia, when it is devoted to a task like this and so triumphantly achieves its purpose. Trotsky as Commissar of War at this moment might hardly add much to his political record; while with this book he opens up another field of influence to himself — a field that is often more important than a man's achievement in actual politics. Bacon, as the author of the *Novum Organum* and the *New Atlantis*, was better occupied than as Lord Chancellor, though he did not appear to think so; and it is to political exile that we owe the *History of the Peloponnesian War* and Clarendon's *History of the Great Rebellion*, no less than *Paradise Lost* to political defeat.

It is more difficult to define what *kind* of a book this is. Mr. Max Eastman, its translator, claims that "this is the first time the scientific history of a great event has been written by a man who played a dominant part in it". There is no doubt about the "dominance", but more doubt about the "science". For the real claim of this book is not that it is an impersonal, a scientific history; though, indeed, it is a brilliant example of a very rare species, a history that is inspired by the conception of society and the forces at work in it, implied by historical materialism. This, in short, is *a* Marxist history, but not *the* Marxist history of the Revolution; for that we shall have to wait for some future Pokrovsky, altogether more impersonal,

[1] *The History of the Russian Revolution*, by Leon Trotsky. 3 vols.

An Epic of Revolution: Reflections on Trotsky's History

more objective; but, no doubt, that will be a much duller affair. Whereas this is alive and tingling in every nerve. It has all the brilliant qualities, and the defects, of its author's personality. It has extreme definiteness of outline, a relentlessness towards his enemies that goes with it, dramatic sense and visual power, a remarkable sympathy for the moods of the masses with a gift for vividly portraying them — the qualities we should expect from a great orator; and, in addition, the political understanding of a first-rate political figure.

It was noted by Macaulay how incomparably superior in the understanding of politics any political pamphlet of Swift's was to one of Johnson's; just because of the intimate contact of the former with politics, and hence his correct judgment of the forces that as a writer he was estimating. The same holds good of Trotsky as an historian. He may not be impartial (neither for that matter is anybody else); but what a political grasp is revealed on every page, in the chance remarks thrown out as he proceeds, compared with the laborious irrelevancies of academic historians. Louis Madelin, whom Trotsky tilts against on the subject of impartiality, is hardly worthy of his notice; but his contempt for those historians who confuse the symbol with the interests behind it, is both salutary and generally applicable. Are there not historians who think that the Bourbons failed to re-establish the monarchy in France after 1870 just because of Henry V's attachment to a little white flag? Or historians of a party who write on the Tory or Whig parties without ever realising what a party is when you analyse it?

Not so Trotsky. He has an illuminating comment on the curious confusion of the July days, when both the Bolshevik insurrectionaries and the loyal troops of the Government wanted to submit to the Executive Committee of the Soviets as their sovereign authority. The fact was that the Committee represented both the petty-bourgeois and the workers, and the troops recognised it as the representative of the one, and the workers as the representative of the other. Two conflicting class interests were bound up in the same institution; it was symptomatic of the character of the Dual Power, and only time could resolve the conflict. But the fact that there was only one symbol, the Committee, did not mean that there was nothing to fight about; it only papered over the cracks. Trotsky comments:

If this conflict had taken place towards the end of the Middle Ages, both sides in slaughtering each other would have cited the same texts from the Bible. Formalist historians would afterwards have come to the conclusion that they were fighting about the correct interpretation of texts. The craftsmen and illiterate peasants of the Middle Ages had a strange passion, as is well known, for allowing themselves to be killed in the cause of philological subtleties in the Revelation of St. John, just as the Russian separatists submitted to extermination in order to decide the question whether one should cross himself with two fingers or three. In reality there lies hidden under such symbolic formulae — in the Middle Ages no less than now — a conflict of life interests which we must learn to uncover. The very same verse of the Evangelist meant serfdom for some, freedom for others.

He makes the point clearer in the case of the June days in Paris in 1848, when the cry *Vive la République* went up on both sides of the barricades:

To the petty-bourgeois idealist, therefore, the June conflict has seemed a misunderstanding caused by the inattention of one side, the hot-headedness of the other. In reality the bourgeoisie wanted a republic for themselves, the workers a republic for everybody. Political slogans serve oftener to disguise interests than to call them by name.

It was impossible to expect Trotsky to suppress his own personality in the book; not only for the reason that he is Trotsky, but because, after all, he played such an important part in the Revolution. To have suppressed himself would be a falsification of history. But he does go much further towards impersonality than one would have thought possible from one of his temperament. He writes throughout in the third person; he keeps himself in the background of the picture. The book gives an impression of a highly exciting personality, but not one of egoism; and, with one notable exception, it leaves an impression of fairness, at least not of unfairness. In

An Epic of Revolution: Reflections on Trotsky's History

the light of events he seems justified in his merciless characterisation of the Tsar and Tsarina, Miliukov, Kornilov, Kerensky, and many of the Socialists. The exception is, of course, Stalin.

It is a pity that his personal feud with Stalin has prevented him from recognising Stalin's part in the Revolution. Whenever he comes near the subject, the history tends to turn into a political pamphlet; and one is tempted to think that Trotsky writes history, as the celebrated Dr. Clifford was said to offer extemporary prayer, for the purpose of scarifying his enemies. Nobody would guess from his account that in the October Revolution, though Trotsky was the President of the Military Revolutionary Committee of the Soviet, which organised the insurrection, Stalin was responsible for the organisation of the Bolshevik Party, apart from the Soviet in which other parties were included, to the same end. Over the struggle within the party in October, when Lenin was forcing them into insurrection and the party was divided in opinion, it seems needless to attack Stalin, as the editor of *Pravda*, for trying to tone down the differences: it is the function of a party organ to gloss over the differences within the party, before the eyes of the outside world. Nor, though Trotsky allows that Stalin's defects are not due to lack of character, as in the case of Kamenev and Zinoviev, the two opponents of the insurrection, is it reasonable to attack him on the ground of his caution. There are leaders and leaders. It is true that Stalin is not of the tempestuous, romantic type of revolutionary like Trotsky, but he is none the less a great leader. He reminds one rather of Burghley in our own history, who had a great gift for taking cover. But that did not prevent him from being bold and courageous in policy, as in the case of the great leap in the dark of 1559, when this country was committed finally and decisively to the Protestant Reformation. And so, too, Stalin is the man, after all, who has taken the plunge of committing Russia to the Five Years' Plan.

That said, one can pay tribute to the extraordinary and original qualities of the work. It is a kind of prose-epic of the Revolution. The Revolution is Trotsky's hero, and he displays an attitude of reverence towards it, very proper when one

The End of an Epoch

considers the vastness of the subject and the profundity of its issues, and singularly appropriate in so irreverent a character as the author. It is as if the Revolution were the one great revelation of his life, a kind of beatific vision which it must be very exasperating to be withdrawn from, even such a distance as Prinkipo is from Odessa. His sympathy with the masses, remarkable in so impatient a person, is surprising in its reality and imaginativeness. It is easy enough when, as in the February Revolution, the masses are on the up grade, feeling their way forward, carrying the soldiers with them; he picks out vivid, significant little episodes of the " struggle of the workers for the soul of the soldiers ", the fraternising of the men from the factories of the Vyborg district with the old and famous regiments quartered in the capital. But in the more difficult days of July, when the Bolsheviks were defeated and Lenin driven into hiding, when the workers and soldiers were stupid enough to believe the slanders that were sedulously propagated by the Provisional Government to the effect that Lenin and Trotsky were German agents, he still defends them. He argues that it was only a superficial change in the minds of the people; that if they " really did change their feelings and thoughts under the influence of accidental circumstances, then that mighty obedience to natural law which characterises the development of great revolutions would be inexplicable ". That, he implies with a superb rationalism, is quite inadmissible; in one place, he speaks of the " insulted reason of history " — a phrase so revealing in its optimism. The deeper the masses are caught up in the revolutionary process, he claims, " the more confidently you can predict the sequence of its further stages ". Only — and he gets out of the difficulty by adding — " you must remember that the political development of the masses proceeds not in the direct line, but in a complicated curve ".

Yet one wonders a little whether he does not over-estimate the intelligence and the willed activity of the masses as such; whether he does not dramatise them too much, having seen them in action when stirred up from the depths and at the top of their form? But this has its good side; there is no facile

An Epic of Revolution: Reflections on Trotsky's History

scorn of the people; he calls even scepticism with regard to their latent potentialities for action, "cheap". And he is fundamentally right: he remembers the source of their ineptitudes and mistakes, when one is tempted to think of their consequences in their own suffering. Even when the dregs of the population, the criminal element, see their chance and come out of their holes, he says a little pityingly, Here is the barbarism that the barbarism of the old order created! When one considers the horror of a régime run by a Rasputin, the weak and cold-blooded Byzantinism of the Court, the criminal levity of the aristocracy, the disasters, the corruption, the repression, was not the Tsardom but a "crowned hooliganism"? Nor is the judgment of the masses so blind and credulous as is so commonly supposed, he says. They have a shrewd idea of the factors affecting their own action, he remarks concerning the conviction of the people that the responsibility for the July clashes was that of patriotic provocateurs. "Where it is touched to the quick, it gathers facts and conjectures with a thousand eyes and ears, tests rumours by its own experience, selects some and rejects others. Where versions touching a mass movement are contradictory, those appropriated by the mass itself are nearest to the truth." He goes on to pour scorn on those historians typified by Taine, who, "in studying great popular movements ignore the voices of the street, and spend their time carefully collecting and sifting the empty gossip produced in drawing-rooms by moods of isolation and fear". True enough; but the moods and thoughts of the people themselves are the most difficult of all historical material to collect; up to now, theirs has been a silent contribution, though their fate is the burden of history. Will the Revolution bring freedom and consciousness to them? Trotsky, with the Communists, believes so: it is for them, the whole justification for Revolution, the reward of intolerable effort and no less intolerable suffering — the end that he calls "*the spiritualisation of the inert mass*".

It must be said that artistically the book is completely successful in developing its theme, the gradual self-realisation of the people under the pressure of revolutionary circumstances.

The End of an Epoch

It is as if one watches the people coming alive, thawing after the long winter of repression, shaking themselves free from the unmeaning and frozen gestures of the ages. It is a progress from the old Russia, when on the outbreak of war the Tsar went to find strength in prayer at the shrines of the Kremlin, and the people kissed the ground he trod on; when the War Minister declared to the Duma, as the whole front was caving in, in 1915, " I place my trust in the impenetrable spaces, impassable mud, and the mercy of St. Nicholas Mirlikisky, Protector of Holy Russia "; a progress to the dream that inspired the whole lives of Marx and Lenin, that all these millions of beings might realise themselves as concrete conscious selves.

Intellectually, it is no less successful in its formulations, though here there is more scope for variety of interpretation and some dispute. He analyses the Revolution down to its roots in the backwardness of Russian economic development as a whole, cheek-by-jowl with the most advanced and concentrated industrialism in and around Petrograd and Moscow, the two great revolutionary centres. We all know the primitive agricultural conditions that prevailed so largely in Russia, and that form at once such an obstacle to Communism and such an objective to overcome. But one little realises the extraordinary concentration of the industrial proletariat; of a total of two millions, some half a million were concentrated in Petrograd, in a hundred and forty giant factories, of which the Putilov factory with forty thousand workers was the type. This proletariat had had from 1905 a long revolutionary experience and was constantly in the forefront of political action and engaged in strikes and lock-outs. Of the land, besides the enormous properties of the Crown and the Church, the great landlords owned a quarter. The whole stretch of arable land within the limits of European Russia was estimated at 280 million dessiatines;[1] of this, seventy million dessiatines belonged to 30,000 great landlords. As Trotsky tersely puts it: " This seventy million was about what would have belonged to ten million peasant families. These land statistics constitute the finished programme of a peasant war."

[1] A dessiatine = 2·7 acres.

An Epic of Revolution: Reflections on Trotsky's History

On the top of this economic substructure — in itself unstable, since there was no strong middle class to bind the whole thing together, and the major portion of the finance and industrial capital was held by foreign capitalists — there was a ramshackle political superstructure: a decadent aristocracy, at the head of which stood an exhausted and worn-out autocracy, the whole thing riven by the struggles of the past twenty years. After three years of the war, years of unmitigated disaster, everybody wanted to get rid of the hopeless Nicholas II and his *entourage*, Tsarina, Rasputin, and all; even the aristocracy was in favour of a Palace Revolution — as the genial Protopopoff admits, the very highest classes became *frondeurs* before the Revolution. The only question was, who was to make the Revolution and for whose benefit? The impossibility of settling this question between the aristocracy and the bourgeoisie might have meant the continuance of the Tsardom to this day if the people had not stepped in. It was the people who took the issue into their own hands with the February Revolution; but owing to their inexperience they allowed the power they had won to slip into the hands of the bourgeoisie, who used it to continue for their own benefit substantially the same system: the war, a foreign policy of annexation, capitalism in industry and the great estates on the land. It was this that necessitated the October Revolution before power could come into the hands of the workers, to be used for their benefit. The history of the intervening period, between February and October, is one of utter governmental and social instability, with power swaying now to this side, now to that, the Provisional Government now leaning on the bourgeoisie, now on the workers, staggering from crisis to crisis, until at length the Bolsheviks put a blessed end to it by bringing power to rest broad-based upon the workers.

One does not need to halt over Trotsky's characterisation of the old régime; like the image of the stricken tree, in which Swift saw the premonition of his own fate, it was dying from the top downwards. What is more curious to observe is that Trotsky, who like all Communists would deny the primary importance of ethical motives, is really shocked

The End of an Epoch

by the putrescence at the top. The Emperor, who was unmoved when hundreds of people were crushed to death at his coronation, when thousands were shot down unarmed in the streets in 1905, but who wrote on a report that a certain officer was executing soldiers without any trial, " Ah, what a fine fellow " ; the Empress, who right up to the end, against the remonstrances of all her own family, was pressing the Tsar for extreme measures: " Anything but this responsible ministry about which everybody has gone crazy. People want to feel your hand. How long they have been saying to me for whole years, the same thing: Russia loves to feel the whip. That is *their* nature." Of the one, Trotsky writes: " This orthodox Hessian, with a Windsor upbringing and a Byzantine crown on her head, not only ' incarnates ' the Russian soul, but also organically despises it. This German woman adopted with a kind of cold fury all the traditions and romances of Russian medievalism, the most meagre and crude of all medievalisms, in that very period when the people were making mighty efforts to free themselves of it." Of the Tsar, he says: " This ' charmer ', without will, without aim, without imagination, was more awful than all the tyrants of ancient and modern history ".

The real importance of Trotsky's *History* does not lie in his power of word-painting, either of character or scene; though indeed his gift is so brilliant and incisive that one is continually reminded of Carlyle. There is something of the same technique, the same mannerism even, in the way the rapid lights shift across the scene and particular odd episodes are brought out in singular sharpness of relief and made to bear general significance; something of the same difficulty in following the sequence of events — the lights are so blinding — one may add. But where Carlyle had but his magnificent powers of intuition to rely on, Trotsky has a theory of history at his command, which enables him to grasp what is significant and to relate things together. The same point can be illustrated more appositely by comparison with Winston Churchill's *The World Crisis*, for the two men are not dissimilar in character and gifts of mind. But here again one notices the difference; for Mr. Churchill's

An Epic of Revolution: Reflections on Trotsky's History

History, for all its personality, its vividness and vitality — points which it has in common with Trotsky — has not a philosophy of history behind it.

What distinguishes this work is the basic attempt Trotsky makes to define his subject, to make clear the methods appropriate to understanding it, and to see it in relation to the whole historical perspective opened up by Marxism. He states in his Preface: " The most indubitable feature of a revolution is the direct interference of the masses in historic events ". He explains how, in normal times, when society is not shaken up from its foundations, the events that happen on the surface of society, more or less in political circles, come to be identified with the history of the whole society. A case of mistaken identity that is all very well in quiet times when the surface is not too unrepresentative of what is going on beneath, but is entirely inadequate in time of revolution, when the old order has become too atrophied for the forces that are boiling up underneath, and the masses break through its framework to create a new régime. " The history of a revolution is, then, first of all a history of the forcible entrance of the masses into the realm of rulership over their own destiny."

Trotsky explains that the changes that are brought about in the economic bases of society and in the social substratum of classes in the course of the revolution are not enough alone to explain its creative activity. There is nothing mechanical or schematic about the course of revolution: " The dynamic of revolutionary events is *directly* determined by swift, intense and passionate changes in the psychology of classes which have already formed themselves before the revolution ". On the other hand, neither is the process a spontaneous one, a sort of spontaneous combustion, arising for no known reason out of no discernible circumstances, as liberal idealism would give one to suppose. No; the revolution has its own logic, which is part and parcel of the logic of history, dynamic, living, flexible, capable of innumerable variants, but not without reason. The swift changes in the views and moods of the masses, so characteristic of revolutionary epochs, do not take place in the void, nor are they unrelated to the economic and social conditions that provide the field in which they move and which limits their effective variation.

The Marxist view of history provides a satisfactory correlation of these two elements — a correlation so close as to form one many-sided but homogeneous process. The danger lies in separating out the various strands in the one process, for the purposes of historical exposition. Nor has Trotsky entirely escaped criticism from other Marxists on this account. He sees the danger in the purely psychological approach, " which looks upon the nature of events as an inter-weaving of the free activities of separate individuals and their groupings " ; and describes his own approach as that of

the materialist method which disciplines the historian, compelling him to take his departure from the weighty facts of the social structure. For us the fundamental forces of the historic process are classes ; political parties rest upon them ; ideas and slogans emerge as the small change of objective interests. The whole course of the investigation proceeds from the objective to the subjective, from the social to the individual, from the fundamental to the incidental.

In the course of his treatment of the Revolution, his activism of temperament, perhaps also the fact that he was a participant in the action, leads him to put a stress upon the psychological factors that laid him open to the criticism of Pokrovsky on the score of idealism. Trotsky writes :

The immediate causes of the events of a revolution are changes in the state of mind of the conflicting classes. The material relations of society merely define the channel within which these processes take place. . . . It is impossible to understand the real significance of a political party or find your way among the manœuvres of the leaders, without searching out the deep molecular process in the mind of the mass.

Does not this give the process too idealist a cast of character ? Granted that the historian must search out the molecular changes and adaptations going on in the minds of the masses — since these things are happening to men, not blocks of wood ; yet the changes in the course of events are determined by changes in the external situation — by the economic collapse

An Epic of Revolution: Reflections on Trotsky's History

between February and October, for instance, by the shortage of food, by the impossibility of carrying on the government on the old social foundations, *i.e.* by the inner contradictions of the existing order. This is the ground of Pokrovsky's charge, who looks for the motive force of the Revolution to the objective shifts in society rather than to the mental processes of the masses.

The real problem is how to correlate the two; and it may be that Trotsky and Pokrovsky are not in such direct opposition as they supposed, that their views are complementary rather than antithetical, and that much of their controversy is simply due to their being at cross-purposes with one another. It is only to be expected that Trotsky, as an active participator in the events, would be altogether more activist in his sympathies and his interpretation than the Professor. The danger on the side of the Professor would be that of under-estimating the part played by the conscious and willed activity of the masses — a tinge, perhaps, of fatalism. Nevertheless, one is inclined to think that the Professor sees the whole thing in the better and more balanced historical perspective; Trotsky's book is above all things a guide to action; the Professor is more concerned with adding to knowledge. Yet it is a great strength of Trotsky's book that he never forgets that the figures behind these great events are human beings. " Let us not forget ", he says, " that revolutions are accomplished through people, although they be nameless. Materialism does not ignore the feeling, thinking, acting man, but explains him."

This very activism is a great asset when he comes to estimating the role of parties in the Revolution, in particular that of the Bolshevik Party, and the importance of Lenin's leadership in it. He remarks at the outset:

> Only on the basis of a study of political processes in the masses themselves, can we understand the role of parties and leaders, whom we least of all are inclined to ignore. They constitute not an independent, but nevertheless a very important, element in the process. Without a guiding organisation the masses would dissipate like steam not enclosed in a piston-box. But nevertheless what moves things is not the piston or the box, but the steam.

The End of an Epoch

Later on, when he comes to the crucial question of what was Lenin's contribution to the Revolution, how it would have developed without his leadership, Trotsky's view wins our complete agreement; the question itself provides a kind of test of the satisfactoriness or not of his general historical conception. He emerges from it triumphantly, in one of the most memorable passages of the book (vol. i, pp. 341-2); though, indeed, it is only a vulgar misconception of historical materialism — alas, one that is all too common — that supposes it to exclude all possibility of individuals influencing the course of events. Of course, they have some scope of influence; the point is to know *how far* they may influence events. They cannot transcend the field of conditions within which they are acting; they do not start with a clean slate, as historical idealists suppose, upon which they may write anything; they are bound and limited at every turn by the forces that exist in the field of their action. But to understand this, is itself the greatest aid to effective action that there can be. It was the fact that Lenin understood what was possible on this basis, that his whole mentality was guided by this outlook and method, that made him incomparably superior to any other leading figure in the field.

One does not need to compare him with the Miliukovs and Kerenskys, there is a more useful comparison to make with the British ambassador, Buchanan, by no means an unimportant actor on the same scene. Buchanan, after a lifetime dealing with the surface events of history, as a diplomat, made an heroic effort to accommodate himself to the circumstances of a revolution; but one finds him forced to confess in mid-stream, " The situation is so obscure that I personally see no daylight " [1] — an utterance that would be quite inconceivable for Lenin. Or again, Buchanan's comment on the victory of the Bolsheviks in October: " The inability of Russians to work cordially together, even when the fate of their country is at stake, amounts almost to a national defect ".[2]

The case was quite otherwise with Lenin, not so much because he had lived his whole life in the atmosphere of revolu-

[1] *My Mission to Russia*, vol. ii, p. 160.   [2] *Ibid.* p. 217.

An Epic of Revolution: Reflections on Trotsky's History

tion, but because in Marxism he possessed a guide to the deeper shifts in society, and not merely the surface ripples of diplomatic circles, ambassadorial lunches, state banquets, the quarrels of crowned heads, the idiosyncrasies of Empresses. The masses are more important.

The chief strength of Lenin lay in his understanding the inner logic of the movement and guiding his policy by it. He did not impose his plan on the masses; he helped the masses to recognise and realise their own plan. When Lenin reduced all the problems of the Revolution to one — " patiently explain " — that meant that it was necessary to bring the consciousness of the masses into correspondence with that situation into which the historic process had driven them.

How, then, would the Revolution have developed if Lenin had not reached Russia in 1917? Trotsky has no difficulty in allowing that the role of Lenin was one of decisive importance at that time; it was a case of the perfect conjunction of circumstances and the man. There is no question of the man creating possibilities that were not given in the situation; the situation itself was ripening for further stages in the development of the Revolution, and the leader and the party to take advantage of it were there. Trotsky sums up:

If our exposition proves anything at all, it proves that Lenin was not a demiurge of the revolutionary process, that he merely entered into a chain of historic forces. But he was a great link in that chain. The dictatorship of the proletariat was to be inferred from the whole situation, but it had still to be established. It could not be established without a party. The party could fulfil its mission only after understanding it. For that Lenin was needed.

But without him, is it possible to say with certainty that the party would have found its road?

We would by no means make bold to say that. The factor of time is decisive here, and it is difficult in retrospect to tell time historically. Dialectic materialism has nothing in common with fatalism. Without Lenin the crisis, which the opportunist leadership was inevitably bound to produce, would have assumed an extraordinarily sharp and protracted character. The conditions of war and revolution, however, would not allow the party a long period for fulfilling its mission. Thus it is by no means excluded that a disorientated and split party might have let slip the revolutionary opportunity for many years. The role of personality arises before us here on a truly gigantic scale. It is necessary only to understand that role correctly, taking personality as a link in the historic chain.

It is no part of our task, even if it were possible, to sum up Lenin's part in the Revolution. It is sufficient to say that the impression that arises from the wealth of detail with which his activity is treated here, is absolutely at one with all that we know of him from other sources. There is no more *consistent* character in modern history than Lenin; he is all of a piece, without a shadow of equivocation or question or doubt, with not a trace of the duplicity so necessary to the existence of ordinary leaders, utterly simple, unconscious of himself, completely devoted to one end. Of all the stories of him, one that is most revealing is that told by John Reed in his *Ten Days that Shook the World*; how in the crisis of the October Revolution, Lenin appeared, after months in hiding, to the Soviet Congress which under his guidance was constructing the dictatorship of the proletariat. Most of the deputies there had never set eyes on him, though his name had gone all over Russia; and when he appeared, there was a storm of cheering that lasted for minutes, at the end of which, quite unmoved, he said: " We will now proceed to construct the Socialist order ".

There proceeded from that Congress, what might never have emerged without him, the three great decrees in which the Soviet régime announced its mission: the appeal to all the peoples to make peace on the basis of no annexations; the declaration of all the land as national property; the establishment of the rule of the working class. One is tempted to think there is something very un-Russian about Lenin: the intense practicability, the concentration upon action, his absorbing concern with what is possible and his insistence upon the neatest, shortest way of doing it. It is so unlike Zinoviev, all over the place, or Kamenev, never coming to the point, or so many other Russians without sense, as somebody said, of time or space. Lenin might be a really first-class Englishman (not Mr. Baldwin's sort of Englishman) if it were not for his inexcusable concern with theory, and his (to us incomprehensible) determination to make action and his intellectual position march together.

And yet, for all Lenin's devotion, and all his genius, nothing of the October Revolution might have come about when it did, if it had not been for a peculiarly fortunate concatenation of circumstances. It is a sobering, an ironical thought, to think that he might have gone on spending the last years of his life in exile, like the earlier ones, seeing in his own lifetime no fruit for his efforts. For, as Trotsky admits, " a coincidence of all the conditions necessary to a victorious and stable proletarian revolution has so far occurred but once in history: in Russia in October 1917 ". If this is so, what is the point of the Communist complaints against German Social Democracy for not bringing off the Revolution in Germany in 1918? Lenin himself has said that the necessary condition for the seizure of power by the proletariat is " to have at the decisive moment, at the decisive point, an overwhelming superiority of force ". The Bolsheviks had this in Petrograd in 1917; but have the workers in any Western country, England or France or Germany, ever had a favourable situation for successful revolution? If not, then why the constant Communist recriminations against the failure of Western Socialism to have achieved revolution? These questions verge upon political issues; suffice it to say that the Communist position with regard to them can be met, and so far as I am concerned, will be dealt with elsewhere.

Trotsky's great work draws to an end on a subdued note in which it is not difficult to detect a certain scepticism, or perhaps one should say, philosophical acceptance of the world's ways. Was it, after all, worth all the sacrifice? something seems to whisper in the intermittences of the heart. His first attempt to answer these promptings does not satisfy. " It would be as well to ask in face of the difficulties and griefs of personal existence: Is it worth while to be born? " These melancholy reflections are, he thinks, in general, unimportant. Perhaps so; but they may be important for certain exceptional people — as they must have been for Lenin himself at some stage of his life, before his decision was irrevocably made. But in general, since these things have to be, the only reasonable course is to direct them into the best channels and to minimise the suffering they involve. The suffering involved might have been greater if October had never been; and it need never have been so great if other forces had not prolonged the struggle needlessly against the solution that history itself indicated. For these forces had only a dead end, a blind alley, to offer; whereas, through the October Revolution, the future was made possible, and not for Russia alone.

That seems to be the answer Trotsky and the Communists would make. Nor can we doubt that the October Revolution is the most important world event of our time; it has the significance for us that the French Revolution had for the nineteenth century. It is evidently necessary to know where we stand in relation to it; and to know this we have to understand it. In spite of the deluge of literature on the Revolution produced to delude fools, there is now a reliable literature growing up, on the basis of which we may judge; and in Trotsky's *History of the Russian Revolution* we have had the unbelievable good fortune of a work worthy of that great event.

XXII

QUESTIONS IN POLITICAL THEORY

(1) [1941]

THE foundations of conventional and academic political thinking are, and should be, along with much else, in the melting-pot. For, as Professor Carr reminds us, political thought is itself a form of political action; and, as we have reason to remind ourselves, so many of the categories and modes of that tradition of thought we were brought up in have ceased to mean anything much, a stock-in-trade for which the contemporary world has no use. Professor Carr's book is a suggestive attempt to think out anew the nature of politics, particularly on the international side, in terms that do correspond to the forces at work — in the way the old did not — and that do mean something. His book has been out some time now: all the better opportunity to see how his views and theories stand the test of circumstances.[1]

Professor Carr is a realist, not, as some would think, a cynic about politics. He places himself under the wing of Bacon, whom he cites: " On account of the pernicious and inveterate habit of dwelling on abstractions, it is safer to begin and raise the sciences from those foundations which have relation to practice ". He sees, as against the idealists and liberals who have played such a part in forming the somewhat foggy climate of English opinion, that politics is always and inescapably about power. That is its subject matter. So that the phrase beloved by sentimentalists of the Left, " power politics ", is exceedingly silly — as if politics could be any other! (What is their alternative — " impotence-politics " ? I fear that only too exactly describes it.) On the other hand, he sees, as against the cynics, that politics is not merely about power, but that

[1] *The Twenty Years' Crisis, 1919–1939*, by E. H. Carr.

morals enter in; he describes their relations in that field as an uneasy equilibrium.

About the relevance of Professor Carr's theoretical approach there can be, to my mind, no doubt. The question is whether he goes far enough, whether he has reduced his views to consistency, whether he has worked out their full implications, or left important gaps. Let us get out of the way the obvious, and really accidental, weakness of his position in practical politics: his attachment to Chamberlainism. Again and again he seizes the opportunity to trounce the opponents of Mr. Chamberlain's policy as utopians, and to laud Mr. Chamberlain as a realist. All that reads pretty fatuously now: Professor Carr should take the opportunity to omit these passages from future editions: they only serve to prejudice the reader against the general thesis, to which they are irrelevant. (A professor should also be able to spell Mr. Attlee's name correctly: perhaps the consistent mis-spelling reveals some unconscious bias?)

But these considerations are unimportant compared with the fact that he appreciates fully the social roots of the utopian and realist modes of thought, the social conditioning of all political thinking. In the light of this the problem becomes how far there is any objectivity, any disinterestedness in political thinking at all. Is not this what Pareto was concerned with? Though Mr. Carr may not have read Pareto, he has been greatly influenced by Marxist thought — himself not a Marxist — and this has led to some of the best sections in the book, such as the chapter on the " Harmony of Interests ". He shows how the doctrine acquired a new force in the nineteenth century and " became the ideology of a dominant group concerned to maintain its predominance by asserting the identity of its interests with those of the community as a whole "; and that its success and strength were due to the " unparalleled expansion of production, population, and prosperity " in that period. So much for *laissez-faire* as a doctrine for all times and circumstances; in fact it suited the particular circumstances of this country at that time very well. Then he sees, too, the disadvantages attaching to both studies in a rigorous separation of economics from politics; useful for some purposes, it is frustrating in other respects, and he favours a general return to the older conception of " political economy " as sounder and more fruitful.

All this and more besides is excellent. But at the same time there are extraordinary gaps both in the development of his theory, and in his perceptions in the realm of practical politics. What his theoretical system needs is the application at every stage of the concept of general interest, whether within a group, the nation State or the international community, as a criterion by which to test the activities of subordinate groups and discover how far, if at all, their operations are in accordance with it. Then, too, for one who understands so well the strong points of Marxism intellectually and has undergone its influence with advantage, it is curious to find him appreciating so little the importance of the class factor in international politics. He is quite right, as against orthodox Marxists, who have been shockingly obtuse in the matter, to emphasise the national factor, the struggle between nations. But what has given the whole confused character to international politics in the past ten years, what accounts for the paralysis of British and French policy *inter alia*, is the interaction between these two factors, the class struggle crossing the conflicts between nations on the international plane.

In consequence the book displays a complete misunderstanding of the Nazis and their régime. The Professor has his own form of utopianism: he thinks that the satisfied Powers should have made concessions to the dissatisfied. That is probably true enough as a general principle; but for its practical working out it depends entirely to whom the concessions are made and to what effect. The whole evidence over the past decade goes to show that no amount of concession would have " satisfied " the Nazis, and the whole policy of concessionism was a profound psychological mistake that led straight to the war it was designed to avoid. That should open Professor Carr's eyes to the underlying consideration omitted from his system, namely, that there are some régimes to which concessions can be made, and with which cooperation is possible, with others not. It entirely depends on the nature of the political and economic system internally, whether it is possible to cooperate with it externally. Anybody with eyes to see could have told at any time in the past ten years that it was impossible to cooperate with the Nazi régime: its whole dynamic was European and world conquest.

The real problem of international politics becomes, then: on what basis, what kind of economic and social system internally, is international order possible? Democracies, even capitalist democracies, are more peaceable and cooperative than Fascism; they do not reject international order by their very nature. But we need to explore further to discover the social system in which it is possible to root international order, of which, indeed, an international federation is the outward expression of inner character. It is not merely a question of common feeling, as Professor Carr seems to think; one needs to go further on the basis of his own theoretical approach, to find a common power basis, similar political systems resting upon the interests of the great mass of people organised and regulated economically to that end, instead of the attempt, hopeless in its very nature, to found international order in the conflicting interests of ruling classes, capitalists, nation States, etc.

Of all this Professor Carr's last chapter, on " The Prospects of a New International Order ", seems to be unaware: the bedrock foundation is missing; and I am afraid, therefore, that his dedication " To the Makers of the Coming Peace " will equally fail of effect.

(II) [1939]

Professor Robbins has in his brief essay taken a large and complex subject, and it is hardly surprising that it is not more adequately treated.[1] But to the constriction of space there is added a constriction of treatment, in the conception of the subject rather, which it would have been well to avoid. The essay purports " to explain the part played by economic factors

[1] *The Economic Causes of War*, by Lionel Robbins.

in the causation of modern war ". In this it is hardly successful; it is more concerned to refute the Marxist and Socialist interpretation of modern war in terms mainly of economic factors.

It may be agreed that these writers have by no means made the best of their case, that they have tended to over-simplify and exaggerate certain factors (*e.g.* the financial) in the concatenation of forces making for war. But there is here no recognition of the service rendered by these writers in drawing attention to, and aiding our understanding of, this subject of such importance for modern society; nor is there any sign of appreciation of the strength of their case, however inadequately presented.

Professor Robbins investigates a number of cases in which capitalist interests have been involved in the precipitation either of war or a crisis which might have brought about a war. His conclusions, with the typical exception of the Boer War, for he is a Liberal, are, as might have been expected, negative; and, on his view of the case, rightly enough. I am not sure that the Boer War ought not to be excluded too. For the examination rests on an altogether too narrow view of the argument, and of the way to go about examining it. The subject is one of historical causation and the right arrangement of historical factors in that process. Professor Robbins goes into some half a dozen cases. A really satisfactory analysis would have to rest on a great many more, and a more subtle view of historical causation. Any historian could have provided him with more cases — our own commercial wars of the seventeenth and eighteenth centuries, for instance. But that is beside the point.

The point is that the whole concatenation of factors is much more complex; wars are overt conflicts of power, and economic power is fundamental in the social process. The part played by economic factors in the causation of war is, then, more normally an indirect one; but it is none the less effective for that. Let us take the case of the German industrialists who, according to Professor Robbins, were " silly enough " to subscribe to Nazi Party funds. It may be presumed that they knew their own business better than Professor Robbins. Peace and Social Democracy were thoroughly bad for the capitalists of German heavy industry: they were being ruined by it. The Nazi victory saved them. Armaments and the heavy industries boomed, and they prospered. It may be that they were not envisaging war, or deliberately planning it; that only happens as a consequence.

The Marxist and Socialist theory, I take it, is that under the existing system with the immense power groups of monopolistic capitalism, with political and State institutions under their control (instead of the other way round, as Professor Robbins seems to think), with all the conflicts that that gives rise to, war is inevitable. The *occasions* of wars will naturally be political or diplomatic in character, demands upon Serbia, upon Poland, etc. But that does not mean that war is not the consequence of conflicts in social structures of certain kinds, economic in character.

It follows that if the conflicts endemic in these social structures are conducive to wars, then the essential thing is to subjugate them to common order, to bring these powerful, irresponsible groups, the "over-mighty subjects" of the modern State, under public control. Professor Robbins repeats the cliché that Socialist control would only exacerbate such conflicts and make them worse. There is no evidence to show that that would be so. That Professor Robbins's case is indeed a weak one may be seen from the crux of this argument on the "Wars of a National Socialist World" (pp. 94-5) where he refuses to see the distinction between National Socialism — which might more properly be described as National Capitalism — and Socialism. Under the first, group interests are given a monopolistic position at the expense of the community as a whole; Socialism implies public control of such power groups in the interest of the community. The first is by nature national; the second implies international order, for which the large measure of common interest between the working classes of the world provides a foundation. The elimination of the worst conflicts implicit in monopoly capitalism, the subjugation of the search for "private" profit as the motor force to that of the public interest, the removal of that destructive impulse which is the group-appropriation of power in the latest phase of capitalism, these should considerably increase the chances of peace in a more securely organised international order. To slip the argument against National Socialism, where it does apply, and use it against Socialism, where it does not, is rather disingenuous, to say the least.

For what Professor Robbins is looking for is the root of our present discontents, and how to remedy them. He finds it, rightly enough, in the existence of independent sovereign States. (No recognition that this discovery has been a commonplace of Socialist thought since Marx.) Professor Robbins exposes the weak point in his case when he says on his last page: " It was not denied that the existence of independent sovereign states was itself capable of further explanation ". If he explored a little further, he might find the explanation of something more specific, more relevant to his essay, the relation of national State sovereignty to the power factors which in the present phase of capitalism are most operative in causing modern war.

(III) [1939]

Now that the chorus of praise on the appearance of this book has somewhat subsided, perhaps a more critical appreciation, noting its defects as well as its merits, may be in place.[1] Not that I do not admire the book: I think that it is, without exception, the best and the most important that Mr. Russell has written on the subject of politics. It is the most mature. He has been becoming increasingly interested of late years in history, which is the indispensable background for the understanding of politics. That has led to a vast change in him, and one that is wholly to the good. When one remembers the preaching up of the virtues of anarchism in the old, or rather the young, Russell, it is amusing to hear him reminding us of the necessity of government, warning us against the dangers of anarchy. " Men are so little gregarious by nature that anarchy is a constant danger, which kingship has done much to prevent." " There must be a strong government, to prevent crime", etc. " Some subjection of the individual is an

[1] *Power: A New Social Analysis*, by Bertrand Russell.

inevitable consequence of increasing social organisation." It is like hearing a converted Satan rebuking sin. How salutary! But how odd that so brilliant a man should have taken so many years to arrive at what is, after all, only plain common sense.

Nor is the process as yet complete. History has its own discipline and method no less than science; and Mr. Russell's reading of history is too recent for him to have avoided some faulty and some inadequate historical judgments. There is one suggestion of his that we have heard before (I think in *The Prospects of Industrial Civilisation*), that if war is to be prevented, it is likely to be " by an initial despotism of some one nation ". Apart from the reactionary import of this, I should have thought that the whole lesson of modern history is that no one nation can dominate all the rest, and therefore that the true solution for our troubles is some federal international system.

The fact is that Mr. Russell is essentially and always a moralist. In consequence the last chapters of his book, which deal with the ethical questions involved, the moral and political dangers of power and the modes of warding against them, are far the best: there is nothing here that one dissents from, hardly anything that could be bettered.

Even those among us who are least prone to accept the superficial assumptions of democracy at their face value, cannot but be affected by his thesis that modern agglomerations of economic and political power are so irresponsible, so dangerous to their subjects, that they are only tolerable provided they are mitigated by real democratic conditions. " The merits of democracy are negative: it does not insure good government, but it prevents certain evils. . . . All history shows that minorities cannot be trusted to care for the interests of majorities." We can all assent to these propositions and appreciate the value of Mr. Russell's warnings.

It is in the analytical part of the book that he is less adequate. He states his thesis that " the fundamental concept in social science is power, in the same sense in which energy is the fundamental concept in physics "; that power takes various forms, economic, political, military, ideological, and that " no one of these can be regarded as subordinate to any other, and there is no one form from which the others are derivative ". We all recognise that the forms which power takes in society are connected; but the real point to know is *how* they are connected. It is here that a critical understanding of Marxism is of value, and Mr. Russell's prejudice against it, for it is no less, leads him astray and into saying some silly things. There must be somebody to tell the Emperor when he has not got any clothes on.

In a later chapter on " Economic Power ", which with Chapter I is the least convincing in the book, Mr. Russell says that " economic power, unlike military power, is not primary, but derivative "; and he tells us that it is derived from law and public opinion. This is not the case and can be immediately disproved, apart even from the glaring contradiction of this with Chapter I, where he says that the various forms of power are *not* derivative. Which does he mean? The truth, of course, is that if there is one form of power that is less derivative, more fundamental than another, it is economic power. Necessarily so; for people have to eat, gain a livelihood in a certain place in certain conditions before they begin to live well; there is a sense in history in which the economic and social forces set continuously limiting bounds to the psychological and other factors and cannot be transcended. Mr. Russell talks about the love of power being limited by love of ease, etc.; but still more is it limited, and its scope in action determined even, by social factors. However power-loving you might be, if you were a German Socialist, you would not have got through to power, the social conditions being what they were. But a Hitler could. That reflection shows the truth of my contention. For Hitler wanted power on the conditions the anti-working-class forces were prepared to allow him. And he has not fundamentally departed from them, but is rather fulfilling them, even to continuing the pre-war programme of Mittel-Europa and Balkan expansion. That shows how the limiting conditions are set by factors of class and social structure, which are essentially economic in character.

The End of an Epoch

The way in which these factors are inter-connected is very complex; but Russell's view may be directly disproved. He tells us that economic power is derived from law and public opinion, without asking what is behind the law and what makes the opinion. It is very naïve to say that " the oil of the United States belongs to certain companies because they have a legal right to it, and the armed forces of the United States are prepared to enforce the law ". Very naïve indeed, for ultimately the legal system is theirs, and they pay the armed forces. That is putting it crudely, whereas the real situation with regard to force in society is always dressed up a bit, a little papered over and disguised. But the facts of society are crude underneath; and it is not unknown for the great oil companies, when the law is working badly for them, to take the law into their own hands, and employ their own squads of company men to do the necessary.

The whole *tendency* of this book is intellectualist; Mr. Russell is an intellectual — that is why. But the greatest danger for intellectuals in the realm of political thinking, still more in that of political action, is to assume that intellectual considerations matter. It is power that counts. Mr. Russell says: " It is easy to make out a case for the view that opinion is omnipotent, and that all other forms of power are derived from it. Armies are useless unless the soldiers believe in the cause for which they are fighting." There is something in that; but it is a more important consideration that soldiers have to live. That is what matters most.

And so on throughout the first half, the analytical part of the book, which is much less satisfactory than the second half. Even in the treatment of psychological factors, Mr. Russell, like the individualist he is, omits the factors of mass psychology which are more important than those of individual psychology for politics. These are some of the disadvantages, for a writer on politics, in not being an historian. But perhaps a great admiration for the Emperor in so many other respects may excuse a comment upon his occasional nudity.

Questions in Political Theory

(IV)   [1931]

Mr. Tawney's new book is one more contribution, and a very distinguished one, towards a growing body of thought on politics and economics that is coming to occupy the centre of our speculation about society.[1] On its theoretical side, it exemplifies a point of view in which politics and economics are so intimately connected that it becomes clear how useless it is to consider political problems without investigating their economic foundations; it brings into the foreground the question of class, and gives it due importance as a motive force and a formative influence in social relations. On its practical side, and considered as a tract for the times, the book deals with problems of the utmost importance to modern society.

Mr. Tawney's standpoint is a vital and relevant one as to the issues that underlie political conflict. It is no mere academic disquisition as to where sovereignty resides; nor is it, what is scarcely any better, propaganda for some such belated notion as royalism, so ludicrously irrelevant to the time or to what really matters for society. The issues it discusses are real questions of importance: whether the extension of equality would add to or detract from the efficiency of the economic system; whether this equality as a social policy necessitates equality of income; if not, in what directions and by what means equality may be achieved; whether its attainment would react unfavourably upon the standard of culture, that is, whether culture is dependent upon a leisure class; what, in any case, are the moral advantages of such a régime, and the disadvantages of the existing order.

The very effective and biting survey of the "Religion of Inequality" with which the argument opens, quotes a remark, of extreme suggestiveness, out of Arnold's essay on "Equality": "On the one side, inequality harms by pampering; on the other by vulgarising and depressing. A system founded on it is against nature, and, in the long run, breaks down." The suggestion contained in this opens up an interesting train of

[1] *Equality*, by R. H. Tawney.

thought. We hear so much and so often from the protagonists of inequality, that inequality alone is natural to man, that it would have been well to remind them that the struggle towards equality is equally natural. Or, to put it in another way: it is the very purpose of society to counteract the ill effects of natural inequality; but society in so doing is no less natural than is the inequality it seeks to rectify. All human society displays this instinctive tendency to round itself out, to strike a balance by which all its effective members may be provided with a share in subsistence.

This is what the demand for equality means at bottom — an equal share in subsistence. It is a highly abstract conception. But Mr. Tawney recognises, as we all do, that it is no simple thing to be achieved by any single stroke — by Mr. Shaw's cast-iron notion of equalising incomes, for example. It is to be achieved rather by equalising social conditions on one side, by the opportunities society should offer equally to its members; and on the other, by exerting a constant pressure towards equalising endowment, such as income, even though the process may never be complete. It is a complex, many-sided campaign, taking into account the different natural equipments of people, and not necessarily applying the same treatment to everybody. " It involves, in short, a large measure of economic equality — not necessarily, indeed, in respect of the pecuniary incomes of individuals, but of environment, of habits of life, of access to education and the means of civilisation, of security and independence, and of the social consideration which equality in these matters usually carries with it."

Here a word may be interposed to suggest that it is not necessary to the carrying-out of such a policy, as Mr. Tawney seems to suppose, that everybody should receive an equally thorough and stimulating education up to sixteen. Isn't this begging the question? The education individuals should have may need to be very different, to be unequal in fact, in order to achieve a wider, more social equality. It does not need to be the same, either in kind or in length of time, in order to achieve this. The education they are given should be adapted to different equipments of intelligence and aptitude; with some it may be more manual and practical, with others more intellectual and bookish. We should agree that it is only with the proper recognition of equality in society that the latter ceases to be a danger to the former. A man would then be regarded as being no worse for being a bus conductor, and no better for being a bishop. Indeed, earlier in the book Mr. Tawney arrives at the essential and right principle: that the differences society permits should be relevant to the function performed; that is, that the artist needs books and leisure for his occupation, since they are his tools, as much as the machine is of the engineer. He asks only that such gradations as there are " may be based on differences of function and office, may relate only to those aspects of life which are relevant to such differences, and may be compatible with the easy movement of individuals, according to their capacity, from one point on the scale to another ".

The second chapter, and indeed the main body of the book, is concerned with the relation between inequality and social structure. In this part Mr. Tawney makes brilliant use of the material collected by Professors Carr-Saunders, Clay, Ginsberg, and others. But one wishes all the same that he had considered, in the section on " Equality and Culture ", what is to be said on the other side. The problem as he sees it is to end social inequalities without in any way hurting the divergences and variety which are a source of richness to society. And therefore it is necessary to know how far culture is dependent upon this social variousness, and hence upon economic inequality. The defenders of the present system imply, even where they dare not say, that it is so dependent. It may be that a certain desirable variety does come from this juxtaposition of different social classes and standards; the point is whether the advantage, if it exists — and the matter has to be argued out — is outweighed by its disadvantages. Mr. Tawney is clearly of opinion that it is; and much of his book is devoted to expressing what these disadvantages are.

The most original and in some ways the most satisfactory part of the book is the section on " Class ". Here, better than anywhere else in English, may be found a valiant and independent attempt to construct a concept of class. The attempt, in itself, succeeds almost entirely; it grasps, for instance, the essential point, that class is not determined by this or that specific characteristic of a group, but by a totality of conditions all of which must be present to constitute class division. Nevertheless, it is a disadvantage of the moral approach that it is apt to confuse the proper recognition of class and class struggle as facts of society, with the approval of existing class arrangements and their consequences. From this confusion Mr. Tawney is not entirely free; and perhaps because he strongly disapproves of the latter, he ends by underrating the theoretical importance of the former. He deals with the subject with evident reluctance and leaves it with relief; though why the question of class should always receive the tribute of embarrassment, and as often as not of a pained surprise, is amusing to speculate on. But it is not the way to recognise a factor of fundamental importance to society, to regard it as " a sensitive nerve ". Nor is it helpful to describe class struggle as merely " a regrettable incident "; or, if it is, then it is an incident that has continued throughout history.

The same criticism may be made of his treatment of the question of power and its economic connotations at the end of the book. He says categorically : " It is not the case, therefore, as is sometimes suggested, that all forms of power are, in the last resort, economic, for men are so constituted as to desire other than temporal goods and to fear other than economic evils. . . . To destroy it, nothing more is required than to be indifferent to its threats, and to prefer other goods to those which it promises." Of course, men desire other than temporal goods; but there is a point beyond which they cannot choose, or they cease to exist. In this lies the ultimate sovereignty of the economic. If they do choose not to exist, it is none the less a victory for the economic power exerted against them — even if it has no moral victory. And if they should be able to escape this choice, it is in so far as they have some economic power to rely on: the demand for their labour, or the strength of their associates enabling them to hold out. The moral is clear, and it would in no way detract from the cogency of Mr. Tawney's argument to admit it.

However, there is no doubt that he is ready to draw the moral in fact and apply it in practice; even if he is not yet prepared to accept all its implications in the realm of theory.

(v) [1940]

Professor Karl Mannheim is one of the distinguished company of scholars and scientists who have come to us from Germany: a useful recruit to this country. For he is a sociologist, and an eminent one. And it must be confessed that great as the English contribution has been in the realm of politics and political thinking — one has only to recall such names as Hobbes, Locke, Bentham, Mill — we have not as yet made anything like an equal contribution to the scientific study of society, which is what is meant by sociology.

There are reasons for that: one of the most important has been just the stability of our social order. That very security has now become a danger to us, Mannheim reminds us: our social forms and conventions and ideas have gone unquestioned for so long — and it must be admitted that they served us very well in the past — that we are only half alive to the crisis all round us, which is now besieging our shores. We have tended to regard the collapse of liberal democracy, with its decent civilised standards, as " the passing symptoms of a crisis which is confined to a few nations "; whereas " those who live within the danger zone experience this transition as a change in the very structure of society ". Mannheim tells us that those who have first-hand knowledge of the crisis are convinced that " both the social order and the psychology of human beings are changing through and through ", and it is no use, therefore, hoping to remain unaffected by its influence. The best thing we can do is first to understand it in all its aspects, then to take measures to counter its dangers, and to try and plan constructively ahead the future forms of our society and our own adjustments to them.

In all this Professor Mannheim is a first-class, if difficult, guide. He has had the advantage of studying the crisis close up, in the storm-centre of Germany; and for all that we might like to avert our gaze from the more horrid symptoms of it there, we must not overlook the cardinal importance to the modern world of Germany as a crucible of social experiment. Then, too, the Professor is very thoroughly German; he spares us nothing, and his book is written in the most intolerable, abstract, German sociological jargon.[1] The importance of what he has to say is equalled, though not surpassed, by the difficulty of getting at the gist of it. Which makes it the more important to act as a guide to our guide.

Mannheim thinks that " there is only a chance that the western states, with their deep-rooted democratic traditions, will grasp the position in time ", and since he wrote that, one of the greatest of them, France, has suffered a complete collapse, as if to add point to the remark. All the same, he thinks that if we *would* take our opportunities, leap to the greatness of the creative task awaiting us, we have a chance of working out better solutions to the problems of modern society than the imposed solutions of the dictatorships, which are no permanent solutions. If we will not, then the world is open to those who will — however crudely, however unsatisfactorily, and with whatever sacrifice in human suffering and the laboriously achieved standards of civilised life. " At a certain stage of social development ", he says, " it is not enough to leave external trends to themselves; we need a new type of man who can see the right thing to do, and new political groups which will do it."

Here I must put all the cards on the table: I am in complete agreement with Mannheim's general position; I regard the lines he indicates as those along which we must move if we are to come through the crisis afflicting our society with success. Some years ago, in the halcyon, hopeful days of 1929–31, I wrote a book, *Politics and the Younger Generation*, the burden of which was the need for a new and younger leadership. It remains to be seen now, in the middle of a struggle for our very existence, whether that leadership will be grasped, whether the

[1] *Man and Society in an Age of Reconstruction*, by Karl Mannheim.

magnitude of the creative effort needed will be realised, by the new Churchill–Labour Government. So far the signs are distinctly favourable; the curious and interesting thing is that of all our leaders, it is the most unacademic of them, Mr. Bevin, who seems to think instinctively along the lines of this most academic of sociologists.

Professor Mannheim begins with the need for a new psychology of society. I have long realised, as he does, that the psychological assumptions of liberal democracy are completely out of date and untrue to the facts; as a matter of fact, blindly adhering to them in the face of all the evidence pouring in from the contemporary world has been one of the fundamental reasons for democracy's downfall. Liberal democracy assumed the rationality of the great mass of people. I should have thought that what even Mannheim calls the " senselessness " of their behaviour in politics showed how baseless that assumption was. He defines as " substantially rational an act of thought which reveals intelligent insight into the inter-relations of events in a given situation ". In that sense the great majority of people never have been rational about politics — the area is too large for them to grasp and the subject matter too complicated. I should be prepared to go further and say, with Shaw, that the discovery of the bottomless foolery and credulity of people is the prime discovery of the contemporary world. The dictatorships have realised this fact only too well — indeed it is the foundation of their success: they treat the people like idiots and the idiots respond. They know their weakness, their gullibility — Hitler does not disguise for a moment what he thinks of them in *Mein Kampf*, and they exploit it for all they are worth. It is, of course, very wicked, but so far there is not much sign of their peoples giving their masters the lie.

Mannheim is an academic student of all this and would not put it anything like so crudely; indeed he has a great objection to calling a spade a spade. But he knows that it is so all the same, and for all his cumbersome terminology it is fairly clear what he thinks. He analyses the problem very thoroughly and suggests how it has come about. Even the older nineteenth-century Liberals, like Mill, who entertained less illusory views about the faculties of the average citizen, thought that with the extension of industry he would become more rational and capable of political judgment. But Mannheim shows that the industrialisation of our society has by no means led to an increase of " substantial rationality ". In fact it has detracted from it. The whole process of rationalisation of industry has meant ever-increasing specialisation, so that fewer and fewer people see the process as a whole. " A few people can see things more and more clearly over an ever-widening field, while the average man's capacity for rational judgment steadily declines once he has turned over to the organiser the responsibility for making decisions." Actually the old Liberal social order, based upon small capitalism, was a better school for individual reason and judgment.

What this means, then, is that the function of leadership has become of crucial importance in our society, of which a dominant feature is the emergence of the masses into the foreground of politics — hence the stresses and strains of our time. Mannheim is, of course, opposed to the dictatorship solution, and regards it as no solution at all, though it does recognise the problem. Nor is it: since it uses its knowledge of how irrational people are to make them more so than ever, rushes them over the precipice with racial and national slogans that the intelligent among them know to be rubbish. No; the right solution, in the realm of psychology, is the leadership of the irrational by the rational — a rationalism, not superficial and unjustifiably optimistic like that of the nineteenth century, but deepened and fortified by our knowledge (and, alas, experience) of how irrational people are.

In the structure of society this means leadership — not the dragooned, force-imposed leadership of one man, with blind obedience from all the rest. It means the organisation of group leadership, arising out of and always keeping in touch with the mass. The problems arising out of the break-up and reformation of *élites* — to use the rather unpleasant sociological term — are of great complexity and subtlety, and Mannheim has some fascinating things to say on the subject. He lays bare the changes that have taken place in the composition of these *élites*, the breakdown of the old, the function of the intelligentsia. He stresses the importance of not losing the valuable heritage of the older *élites*, but of combining that with the new outlook of the younger ones in the best possible way.

Here — since Mannheim is a new recruit, and in any case prefers generalisation, does not give sufficient concrete examples to help us understand more easily — an Englishman may be permitted to draw attention to the older universities in this country as cases in point. They admirably fulfil this vital function of preserving continuity between the older and the newer *élites*, handing on the standards and the heritage of an older culture to the new. Perhaps this is particularly true of Oxford, which in the years since the last war has done much to develop the social sciences and has thrown open its gates so generously to scholarship boys from all classes of the community — as I can testify, since I am one of them.

The end Mannheim has in view as the proper solution to our difficulties is that of a planned society, based on the appropriate controls and including provision for freedom in the plan. With no illusions, but with a magnificent underlying optimism in face of the discouragements of our time, he speaks of the re-education of the masses and even contemplates the transforming of men's nature. He administers a salutary reproof to those of us who have grown sceptical and cynical through the disappointment of so many of our hopes : a betrayal of the true function of the intellectual he regards that. No doubt he is right, even in regard to that transformation of human nature, since it in part is the expression of our social order and to that extent under our control.

It is impossible to give an idea of the riches of this work, a worthy and a more important successor to Mannheim's *Ideology and Utopia*, which he published in its English form four years ago. This book has a long bibliographical guide to the study of modern society. I prefer to think of the book itself as the guide we want, and in the exact sense, that we need to take it along with us and consult it constantly.

XXIII

MARXISM AND LITERATURE

[1937]

THE last year or two have seen a remarkable movement on the part of the younger writers, poets and critics alike, towards applying Marxist conceptions to literature, and indeed to art in general. It is in this country a very recent tendency. Naturally enough in Russia since 1917 criticism has become largely occupied with these conceptions and literary journals full of the sound and fury of their discussion. Elsewhere on the Continent, too, a good deal more attention has been paid to this approach to literature than with us; M. Marc Ickovicz's *La Littérature à la lumière du matérialisme historique* may be taken as a fair example of the academic treatment of the subject abroad.

But we have lagged behind. Only three years ago Mr. Auden was driven to complain that of the philosophical and the social approach to literature:

I am unacquainted with first-rate work of the latter category. Literary criticisms tend to isolate literature as the relation of one writer to another from the rest of the historical process; their treatment of the effect of the form of a society on art has, as far as I know, only scratched the surface of this profoundly interesting problem.

His contemporaries in the past year have done their best to meet his complaint; and, no doubt because we are late in taking up the subject and feel that we have to make up for lost time, a surprising number of books have appeared on it all at once. There is Mr. Upward's "Sketch for a Marxist Interpretation of Literature" in the symposium *The Mind in Chains*. The Communist writer and novelist, Ralph Fox, completed a book on *The Novel and the People* before he was killed in Spain; his colleague, Mr. T. A. Jackson, has written a book on Dickens from the same point of view. Christopher Caudwell, who like Fox was killed in Spain, left behind him a long book, *Illusion and Reality*, applying Marxism to the study of the sources of poetry. The academic world has contributed Mr. L. C. Knights's *Drama and Society in the Age of Ben Jonson*; the discussion has been carried over into the realm of art, mainly poetry and painting, by Mr. Herbert Read in his *Art and Society*; and the pamphlets and brochures have been innumerable.

It is quite clear that there is a new movement toward in the realm of criticism. The younger writers are turning their back upon purely literary criticism, as other writers are turning away from the purely formal and aesthetic criticism of art. In this, as in so many other regards, they are following up a suggestion, no more than a hint, of Mr. Eliot's. In the re-issue of *The Sacred Wood* he wrote an introduction to say that mere literary criticism was ceasing to have the importance it had for him when he wrote that book; and he went on to suggest that he found it more important in the future to view literature in its historical and social environment, to consider it as an expression of society. So far Mr. Eliot has not followed up the suggestion, unless we are to consider *Murder in the Cathedral* as an essay in historical method — which to some extent it is. But there can be no doubt as to the decisive turn in the minds of the younger writers; and in their new-found concern with the approach to literature in terms of society they are even more unsympathetic to the modern psychological criticism than to the old-fashioned metaphysical school. Mr. Upward, for example, says:

> Metaphysical criticism, though it assumed the existence of eternal spiritual truths, did at least suppose that these truths expressed themselves in terms of the material world, and did therefore attempt to study the material content of literature. The attempt could not be scientific but it led to a fuller and more rational critical theory than modern psychological criticism, with its denial of the material content of literature, can hope to arrive at. The more nearly a critic succeeds in ignoring the objective world, the more limited and irrational will his criticism be.

The End of an Epoch

It is interesting to observe this linking-up of the latest critical tendency with an altogether older-fashioned mode of criticism — a phenomenon that is more commonly noticed in politics than in literature. For the point about the new movement is that it is not so new as it thinks. Mr. Upward describes the basic assumption of Marxist literary criticism to be that " literature reflects and is itself a product of the changing material world of nature and of human society. A poet's images or a novelist's characters are not created out of pure mind-stuff, but are suggested to him by the world in which he lives." This, in that it represents a reaction against criticism purely in terms of literary forms, the consideration of works of art *in vacuo* apart from the conditions in which they take their rise, is a salutary tendency and one that is in keeping with the excellent work that is being done in the historical field. For the latter entirely bears out Mr. Upward's thesis, and the accumulation of facts minutely gathered by modern methods of historical research very much strengthens his case. Such fragments of evidence as Dr. Leslie Hotson has been able to bring together in his *Shakespeare versus Shallow* regarding the Elizabethan Justice of the Peace with whom Shakespeare and his friends were quarrelling, go to strengthen the view that one would have entertained anyhow on general grounds, that Shakespeare was transcribing from his personal experience. The same may be said of Miss F. A. Yates's *John Florio* and her study of the background of *Love's Labour's Lost*, of Professor Sisson's *Lost Plays of Shakespeare's Age* — a book most revealing of the social circumstances in which Elizabethan dramatists wrote their plays. Indeed it may be said that everything that is being brought to light now by these methods of minute research, a small fact here or there brought out of the records in the Public Record Office, a forgotten law-suit in the Court of Chancery, an entry upon the Patent Roll, a will that has been preserved — all this new material goes to substantiate the general view.

But it would evidently surprise these younger critics to know how strongly their view, the historical and social approach to literature, appears in the criticism of a generation ago. The standard history of English poetry, W. J. Courthope's, starts from this view and carries it forward in these terms throughout its many volumes. He says at the outset:

> The very essence of poetry is supposed to lie in the inspiration of the individual poet, the sources of which are beyond the reach of critical investigation. . . . Nevertheless, in all the arts every student soon learns, and every great artist has acknowledged, that those who would excel must take account of conditions which they did not create and can only partially control. . . . The poet is, in a sense, the epitome of the imaginative life of his age and nation; and indeed it may be said that in what may be called his raw materials — his thought, imagination and sentiment — his countrymen cooperate in his work. . . . A great poem is, in fact, an image of national feeling; the inward life of our nation is reflected not less clearly in the course of our poetry than its outward growth in the achievements of its laws, arms and commerce.

This is a statement to which the Communist writer Ralph Fox could whole-heartedly subscribe; nor would he object to its national flavour, for his own book, *The Novel and the People*, displays a very real pride in the creative energy of the English people and the vitality and richness of our cultural heritage.

But Courthope, forty years ago, went further. He saw the importance of class in the history and criticism of literature. His essay on "Poetry and the People" is not less remarkable than Ralph Fox's, or Mr. Upward's.

> All through the history of England [he says] we see a tendency in the life of the nation to concentrate itself in some particular part of the constitution, or in some particular class, which becomes for the time being the sovereign power, rallies round itself all the faculties of the people, represents them to the world, and at last falls into a state of exhaustion.

He suggests how Elizabethan poetry was orientated towards the Court and the Monarchy, expressed the sentiments and standards of that environment and experienced vicissitudes and changes along with it:

> Carry on this idea of the intimate connexion between national life and national poetry into the periods following the Revolution of 1688, and you will see that it helps to explain the changes of taste in the imagination of the people.

The End of an Epoch

He shows how the poetry of the Augustan age expresses the standard of the dominant aristocracy, well placed, well poised, secure in the possession of power: untroubled, and therefore able to concentrate upon formal perfection in their way of life, their intellectual no less than their social manners. He makes the interesting suggestion that " the exaltation of the principle of authority in politics finds its analogy in literature, in the predominant attention paid by the aristocratic writers of the period to correctness of style ". He traces the decline of aristocratic standards in literature along with the aristocracy in politics, and the close correlation between the rise of the middle classes and " the movement in the sphere of taste and imagination which has accompanied it ". He notes, for example, the nineteenth-century departure from poetic tradition, and the " centrifugal, individualising tendencies of the time " which, however much they differed from each other, " agreed in placing the source of poetry solely in the mind of the individual poet, and in disregarding the continuity of tradition in the art itself ". He concludes with a passage that should be very sympathetic to our contemporary *jeunesse littéraire*:

> And hence, now that the middle classes find themselves confronted with collective ideals of order or disorder, opposed to their own notions of individual liberty — the ideal of socialism; the aggressive ideals of other nations; the ideals of the Greater Britain beyond our own islands — they are bewildered and perplexed; and the exhaustion of their capital stock of beliefs reflects itself in the anarchical conflict of ideas embodied in contemporary fiction and poetry.

For an even more remarkable statement of this point of view, we turn to Leslie Stephen's *English Literature and Society in the Eighteenth Century*, published in 1904. It is the most explicit anticipation of this " Marxist " literary criticism that has appeared in this country. For not only does it relate, in the most systematic and satisfactory manner, the content of eighteenth-century literature to the social conditions of the time, but it goes further to apply the conception to literary form.

> We may consider [he says] that any literary form, the drama, the epic poem, the essay, and so forth, is comparable to a species in

natural history. It has, one may say, a certain organic principle which determines the possible modes of development. But the line along which it will actually develop depends upon the character and constitution of the literary class which turns it to account, for the utterance of its own ideas; and depends also upon the correspondence of those ideas with the most vital and powerful intellectual currents of the time.

This last point should recommend itself to Mr. Upward, who is very insistent that for literature to be good it has to be an expression of these vital forces in society; so much so that he says roundly that " no book written *at the present time* can be 'good' unless it is written from a Marxist or near-Marxist viewpoint ". Whether that is so we may inquire later. For the present it is enough to observe that Leslie Stephen, though making a statement that every Marxist would agree with, committed himself to no such dogmatism with regard to the present or future of literature. Perhaps his historical scepticism made him wiser. It is probable that there is now a considerable measure of agreement with Leslie Stephen's statement of the position. The content of literature, the subject matter of poems, novels, plays, is largely given by the social environment; literary forms have an internal evolution of their own, in accordance with their own logic, so to say, but also conditioned by the changes going on in society.

There is a further, altogether subtler and more difficult question on which hardly any light has as yet been thrown by critics, Marxist or other: how far social content determines form in the narrower sense, verse-technique, for example, or the very rhythm and phrasing of words. The influence of a given social environment upon technique in the arts, and again of technical developments upon the arts themselves, is a vast and as yet unexplored field. Take the heroic couplet, for example. May it not be said that something of the very character of eighteenth-century society — a society aristocratic, balanced, with great poise and rational deliberation in its movement — finds its expression in the heroic couplet?

> Two principles in human nature reign;
> Self-love, to urge, and Reason, to restrain;

The End of an Epoch

>   Nor this a good, nor that a bad we call,
>   Each works its end, to move or govern all:
>   And to their proper operation still,
>   Ascribe all good; to their improper, ill.
>   Self-love, the spring of motion, acts the soul;
>   Reason's comparing balance rules the whole.

It is difficult to conceive of that attitude expressing itself in Elizabethan blank verse, or the impulsive rhythms of Shelley, or indeed in any other form than the heroic couplet.

Or consider the fragmentariness, the discontinuity of some contemporary experience, in the broken verse-forms of Mr. Eliot:

>   The river's tent is broken: the last fingers of leaf
>   Clutch and sink into the wet bank. The wind
>   Crosses the brown land, unheard. The nymphs are departed.
>   Sweet Thames, run softly till I end my song.
>   The river bears no empty bottles, sandwich papers,
>   Silk handkerchiefs, cardboard boxes, cigarette ends,
>   Or other testimony of summer nights. The nymphs are departed.
>   And their friends, the loitering heirs of City directors,
>   Departed, have left no addresses.

Or there is the disillusionment that expresses itself in the later, disintegrated blank verse of the Elizabethan dramatists; or the spirit of arrogant confidence of the men of 1588 in the blank verse of Marlowe, the verse-rhythms of Ralegh, as in the prose-rhythms of the letters of Drake. Or again, there is the renewal of confidence and hope in a revolutionary cause expressed in the metrical fibre of Mr. Auden and in the lyrical swing of Mr. Spender's early verse. These things are personal to each writer but they are also social. It is only natural to suppose so. The political movement a man belongs to, the cause he attaches himself to, and the verse-forms he writes in are but different expressions of the fundamental sympathies of his nature; the impulses of the one pass naturally through him to the other. It is a region into which as yet little research has been made.

So far, then, there is much that is valuable in the attitude characteristic of Marxist criticism and with which we can agree. It links on to and continues the distinguished tradition of the historical critics of the late nineteenth century; and for the newer elements in their work it must be allowed that there is a promising field for investigation. Of this, Ralph Fox's *The Novel and the People* is an attractive and a gallant example; it contains much sound sense along with a good deal that is polemical and *parti pris*.

But what of the present and future?

Here these younger critics lead on from a region where they have much to offer, the historical criticism of literature, to one that is much more doubtful, where their influence is likely to be restrictive and even retrograde. There is an extraordinary rigidity in their attitude and in their prescriptions, of which the poets may well beware. They are apt to be so ideological; and their ideology is almost always Puritan. Mr. Fox, for example, hates Romanticism: " that most pernicious form of bourgeois literature", he calls it. Why pernicious? one may ask. The answer is, because Romanticism is apt to be associated with political reaction. That is, the judgment is a political rather than an aesthetic one. And yet what one disapproves of politically may be good art. The history of the appreciation of literature is strewn with such examples. Dryden knew well enough the greatness of Milton, though he cannot have approved of the old doctrinaire Puritan. Marx had the greatest appreciation of Balzac, though the latter from his point of view was a reactionary. Trotsky in his " Literature and Revolution " recognises that Blok was no Bolshevist, but that he was an altogether greater and more significant poet than those who were:

To be sure, Blok was not one of ours, but he reached towards us. And, in doing so, he broke down. But the result of his impulse is the most significant work of our epoch. His poem " The Twelve " will remain for ever.

Lenin remained firm in his attachment to the Russian classics, and had no use for the bright young men who wrote futurist nonsense in the name of the Revolution.

What these things show is that there are aesthetic standards no less than those of social significance, and that Marx and

The End of an Epoch

Lenin and Trotsky tacitly admit that they are of greater importance in judging a work of art as such. Mr. Upward says:

> For the Marxist critic, a good book is one that is true not merely to a temporarily existing situation, but also to the future conditions which are developing within that situation. The greatest books are those which, sensing the forces of the future beneath the surface of the past or present reality, remain true to reality for the longest period of time.

It is all rather vague. What is the future that Shakespeare sensed? or Swift? or Milton? *Paradise Lost* is a political epic idealising the past rather than the future, a magnificent structure of the poetic imagination built up out of the ruins of Milton's outer world as some compensation for defeat. What makes great books last is not only that they are true to the continuum of human relations that subsist through all the changes of differing social environments: love between man and woman, between mother and son, friendship, loyalty of man to man and of man to cause, the cruelty, selfishness, and stupidity of man, and so on. These things are of the nature of man and do not change with social circumstances, though they may be given different colouring and character: this is what gives them lasting value. But equally for a book, a poem, to be good it has to have other qualities as well as social significance: qualities of just proportion and expression, of form and line and phrasing, the significant use of words — in a word, of style.

There are innumerable cases of writers who survive by their style rather than their content; of writers who are read but are not much concerned with the facts of society. The social interest of *Alton Locke* or *Sybil* is much greater than that of *Emma* or *Wuthering Heights*; but the latter are great novels, while the former are not. The reason why they are great would take us too far; but the common sense of literary judgments points to Mr. Upward's standard being, to say the least, inadequate. The fact is that, in judging a particular work *qua* work of art, aesthetic considerations must be decisive. It is in judging a body of literature as a whole, as part of general history, that the social perspective, as Sir Leslie Stephen saw, becomes dominant. And hence it is that Marxism so far has been so much more adequate a method in dealing with the history of literature, than with literary or aesthetic criticism.

That is not to say that the marked tendency on the part of the younger writers to keep closely in touch with the significant movements of our time is not a good thing. Provided that these sympathies are spontaneous and not dogmatically forced, they will prove a fertile source of inspiration: an influence comparable to Mr. Eliot's reintroduction of intellectual content into our poetry. But what is necessary to a satisfactory Marxist criticism is, in addition to developing the historical and social approach, to realise that within this perspective art has a certain autonomy, its own proper standards, and its own inner logic.

INDEX

America, 16, 19, 20, 22, 52, 61, 65, 67, 90-92, 174-5, 268
American Colonies, loss of, 9, 44, 65
American Revolution, 196
Aristotle, 94
Arndt, E. M., 188, 191
Attlee, Rt. Hon. C. R., 11, 69, 107, 108, 112-13, 116-17, 292
Auden, W. H., 316
Augustan Age, 314, 315-16
Australia, 9

Bacon, Francis, 163, 274
Bakunin, M., 191, 261-3
Baldwin, Lord, 3, 9, 11, 14-15, 29, 65-9, 72, 75, 77-89, 105, 110, 111, 118
Balzac, H. de, 317
Barker, Ernest, 135-6
Baynes, Norman H., 209-13
Belgium, 43, 45, 211
Bentham, Jeremy, 168, 172, 173-4
Berlin, 92
Berlin, I., 253-6
Bevan, Edwyn, 25, 30-32, 35-7
Beveridge, Lord, 118, 121
Bevin, Rt. Hon. Ernest, 2, 5, 7, 11, 108, 307
Bismarck, Otto von, 44, 177-9, 184, 186-187, 196, 215
Blok, Alexander, 317
Blum, Léon, 51, 52-4, 56-8, 220
Boer War, the, 43, 295
Boniface, St., 184, 194
Bridges, Sir Edward, 4, 119
British Empire, 47, 197
Brogan, D. W., 217-23
Bryant, Arthur, 81, 83
Bülow, Prince, 177, 197
Burghley, Lord, 42
Butler, R. D'O., 203-6

Cambridge, 153
Capitalism, 148-52, 227-34, 242-4, 295-296
Carlyle, Thomas, 282
Carr, E. H., 261-3, 291-4
Caudwell, Christopher, 311
Cecil, Lord Hugh, 128

Chamberlain, Houston Stewart, 183, 189, 212
Chamberlain, Rt. Hon. Neville, 3, 6, 12-13, 14-15, 24, 29, 65-76, 113, 118, 120, 121, 292
Chamberlainism, phenomenon of, 69-72, 76, 95, 292
Church of England, 97, 125-40; Disestablishment of, 127, 136-40
Churchill, Rt. Hon. Winston S., 2, 5, 6, 9, 10, 12, 13, 15, 24, 37, 55, 61, 66, 69, 73-4, 76-9, 82, 85-9, 122, 144, 210, 282-3
Civil Service, 118-24
Class conflict, theory of, 247-8; interest, 16; and political parties, 153-5, 303-304; middle, and Civil Service, 120-121
Clemenceau, 67
Communism, 28, 32-3, 36, 157, 159, 215, 224-73; and nationalism, 246, 272; *Communist Manifesto*, 226-30; Communists, 21-2, 115-16, 214-15, 220
Conservative Party, 5-6, 8, 9-16, 17, 24, 61, 65-7, 71, 75-80, 83, 85, 94-5, 98, 107, 110
Cornwall, 80
Courthope, W. J., 312-14
Cripps, Rt. Hon. Sir Stafford, 11, 110
Croce, Benedetto, 170-73, 175, 177-8
Cromwell, Oliver, 81, 165
Currency, managed, 147-8
Czechoslovakia, 54, 59, 62-3, 74, 211

Davenant, Charles, 165
Davidson, Archbishop, 125, 126, 130
Democracy, 80-89, 93-102, 294, 305-7
Denmark, 116
Dictatorship of the proletariat, 248, 272
Disraeli, Benjamin, 71, 75, 169
Dryden, John, 317
Dutch, the, 42

Eastman, Max, 274
Economist, 176
Eden, Rt. Hon. Anthony, 25, 66, 113
Education, 101-2, 302-3

321

Eliot, T. S., vi, 224, 236, 311, 316, 319
Elizabeth, Queen, 128, 132
Elizabethan England, 42
Elizabethan poetry, 313, 316
Elliott, Rt. Hon. Walter, 250
Engels, Friedrich, 257-8
English, characteristics of the, 3-4, 39, 181-2, 189, 207, 208, 225, 229, 288
Equality, 301-5
Erasmus, 183, 184, 195
Express, Daily, 2

Fabians, the, 225, 249, 269
Fascism, 26, 28, 32-3, 36, 224, 236, 294
Feuerbach, L., 258
Fichte, J. G., 178-9, 187-8, 195-6
Fisher, H. A. L., 160
Flaubert, Gustave, 222
Foot, Rt. Hon. Isaac, 165
Fox, Ralph, 310, 311, 313, 317
France, 27, 28, 30, 42-3, 45, 51-9, 63, 67, 125, 167, 176, 217-23, 306; the French, 182, 186, 189; French Revolution, 21, 43, 112, 154, 167, 187, 245, 264
France, Anatole, 265

General Strike (1926), 79, 83, 84
Germany, 27, 28, 29, 30, 31, 43-9, 51, 62-4, 79, 87, 91-2, 105, 126, 171, 177-179, 181-216, 306; the Germans, 39, 178-9, 181-216, 223; German propaganda, 39-40, 44
Gladstone, W. E., 169, 177
Government, local, 97-8

Halévy, E., 168
Hegel, G. W. F., 187, 198, 201
Henderson, Sir Nevile, 66
Henry VIII, King, 128, 129, 132
Henson, Bishop H. H., 125, 127, 133, 136, 139
Hitler, Adolf, 11, 24, 26, 40-41, 43-6, 51, 55, 58-64, 74-5, 79, 85, 109, 111, 182, 183, 190-91, 204, 209-13, 299, 307
Hoare, Rt. Hon. Sir Samuel, 66, 69, 72, 86
Holland, 43, 211
Holland House, 174
Hook, Sidney, 253, 256-61
Hotson, Leslie, 312

Ickovicz, M., 310
I.L.P. (Independent Labour Party), 6, 115
India, 246

Inskip, Rt. Hon. Sir T., 85, 130, 138-9
Intellectuals, (mainly Left), 7, 96-7, 112, 155-6
Investment, control of, 148-9
Italy, 29, 54, 111; the Italians, 186, 189

Japan, 28, 51, 61, 90
Jaurès, J., 221, 245, 250
Jefferson, Thomas, 174-5

Keynes, Lord, vi, 1, 2, 83, 141-59, 166
Keyserling, Count, 195
Koestler, Arthur, 20
Krassin, L., 151
Kremlin, the, 19

Labour Government (1924), 237; (1929-31), 50-51, 70; (1945), 21-2
Labour Movement, 95, 99
Labour Party, v, 6, 7, 8-9, 10-14, 30, 103-17, 153-9, 249
Labour Party Conference (1935), 10-11; (1937), 104
Laissez-faire, 142-5, 157
Lansbury, Rt. Hon. George, 6, 10, 11, 105, 110
Laski, H. J., 161-3, 166-9
Lawrence, D. H., 170
Leadership, democratic, 81, 100-101
League of Nations, 34, 49, 55, 113
Lenin, V. I., 235, 237-40, 246, 259, 273, 277-8, 285-9, 317-18
Liberal Party, 108, 115-17, 122, 142-4, 153-5, 160-80
Lincoln, Abraham, 81, 94
Lloyd George, Rt. Hon. D., 17, 66, 77, 78, 144
Locke, John, 166, 168
London, 92, 98, 114
Lords, House of, 97
Lothian, Lord, 25, 32-7
Louis XIV, 27, 30, 43
Low, David, 65, 89
Luther, Martin, 183-4, 195

MacDonald, Rt. Hon. J. R., 29, 84, 221
Machiavelli, N., 161, 163; Machiavellianism, 3, 27, 37, 39-40
Mann, Thomas, 200, 202, 205, 207-9
Mannheim, Karl, 305-9
Marx, Karl, 156, 224-35, 242, 253-67, 297, 317; *Das Kapital* of, 230-35; Marxism, 18, 242, 247-9, 251-67, 283-7, 292, 293, 296-7, 310-19
Materialism, Dialectical, 173, 248-9, 266-7

322

Index

Maurras, Charles, 220
Michelet, Jules, 264
Mill, James, 171, 172, 175
Mill, John Stuart, 156, 165, 168, 172, 232, 308
Milton, John, 274, 317, 318
Moeller van den Bruck, A., 188, 198
Morant, Sir Robert, 121, 123
Morley, Lord, 165, 169, 232
Morrison, Rt. Hon. Herbert, 11, 108, 113-14, 115, 146
Moscow, 102, 267
Munich, 70, 74-6
Mussolini, Benito, 11-12, 26, 51, 54, 55, 58-62, 73, 85, 99, 109

Napoleon, 27, 30, 32, 43, 47-8, 210
National Government, the (1931–40), 9, 10, 15, 29, 30, 33-4, 35-7, 51, 67-9, 71, 75, 84-7, 94, 105, 109, 111, 145
Nazism and Nazis, 21, 63-4, 185, 188-192, 197-8, 208-9, 293-4, 295-6
Nevinson, Henry, 95
New Zealand, 9
Nicholas II, Tsar, 280, 281, 282
Nicolson, Hon. Harold, 70
Nietzsche, F. W., 188, 191
Nonconformity, political, 71-2, 165

Orthodoxy, Left, 7
Orwell, George, 20
Oxford, 101, 153, 309
Oxford, Lord, 158, 169

Paris, 92, 262; Exhibition, 56; June days in, 276
Pelagius, 207
Petrograd, 280
Pokrovsky, M. N., 274, 284-5
Poland, 45, 211, 212, 262
Portugal, 42
Power, 298-300, 304
Prayer Book, the Revised, 126-30, 134, 137
Priestley, J. B., 100
Prussia, 43, 186
Public Concern, the, 146
Public Opinion, 142, 153

Rathenau, W., 200, 201, 202
Rationalist Fallacy, the, 96-7, 142-3, 155, 158-9, 254, 307
Read, Herbert, 311
Rearmament in Great Britain, 13-14, 34, 110-11

Reed, John, 288
Reform Bill, the (1832), 176
Reformation, 161, 164, 184
Relativism, Historical, 258
Religion, Marxist view of, 258-9
Renan, Ernest, 264, 265
Revolution, Puritan, 164-5; of 1688, 166; *v. also sub* America, France, Russia
Ribbentrop, H. von, 92
Ricardo, David, 156, 165, 172, 234
Robbins, Lionel, 294-7
Rockefeller, J. D., 179
Romanticism, 317; German, 260
Roosevelt, President F. D., 53, 91
Rousseau, J. J., 96, 153, 226
Royer-Collard, P. P., 63
Runciman, Lord, 72, 74
Russell, Bertrand, 7, 170-75, 177-80, 242, 297-300
Russell, Lord John, 174
Russia, Soviet, 16-21, 22, 24, 31, 54, 60-63, 74, 157, 310; Russian Revolution, the, 17-21, 112, 236, 237-52, 274-90

Salter, Rt. Hon. Sir Arthur, 121
Santayana, George, 187-8, 192, 196
Scandinavian countries, the, 9, 116
Shakespeare, William, 312, 318
Shaw, Bernard, 7, 170, 271
Shelley, P. B., 316
Shostakhovitch, D., 18
Simon, Lord, 66, 69, 72, 114
Sisson, C. J., 312
Slavs, the, 45, 214
Smith, Adam, 165, 168, 203
Socialism, v, 5, 22, 141-59, 212, 243-4, 296-7; Guild, 146
Spain, 27, 28, 42, 43, 52, 58-60, 62, 73-74, 113, 126
Spengler, Otto, 189-90, 200, 201, 204
Stalin, Joseph, 19, 237, 238, 239-41, 247, 277
Stephen, Sir Leslie, 314-15, 319
Strauss, D. F., 258
Stresemann, G., 91

Tawney, R. H., 301-5
Taylor, A. J. P., 213-16
Times, The, 2, 24-5, 52, 105
Tocqueville, A. de, 176-7, 259
Toynbee, Arnold, 255
Trade Unions, 6, 71, 104, 112, 157, 177, 243

The End of an Epoch

Trotsky, Leon, 237-48, 274-90, 317, 318
Turgeniev, Ivan, 18, 262

United States, 16, 33, 61, 67, 300; *v. also* America
Universities, the older, 309
Upward, Edward, 310-12, 315, 318

Vermeil, E., 193-203
Versailles, Treaty of, 1, 16, 34, 45
Vichy, 223
Victoria, Queen, 174

Webb, Sidney and Beatrice, 268-73
Weitling, Wilhelm, 263
Welles, Sumner, 90-92
Wells, H. G., 7, 93, 152, 159, 170
William II, Kaiser, 45, 191, 197
Wilson, Edmund, 263-8
Wilson, Sir Horace, 4, 73, 119-21, 124
Wilson, Woodrow, 67

York, Archbishop of (C. G. Lang), 136

Zinoviev, G., 237, 277

THE END

PRINTED BY R. & R. CLARK, LTD., EDINBURGH